Principles of
Power Electronics

Principles of Power Electronics

Giani Smith

WILLFORD PRESS

www.willfordpress.com

Published by Willford Press,
118-35 Queens Blvd., Suite 400,
Forest Hills, NY 11375, USA

ISBN: 978-1-64728-327-8

Cataloging-in-Publication Data

Principles of power electronics / Giani Smith.
p. cm.
Includes bibliographical references and index.
ISBN 978-1-64728-327-8
1. Power electronics. 2. Electric power. 3. Electronics. I. Smith, Giani.
TK7881.15 .P75 2022
621.317--dc23

For information on all Willford Press publications
visit our website at www.willfordpress.com

WILLFORD PRESS

Contents

Permissions

Index

Preface

The branch of electronics which seeks to apply solid state electronics for the purpose of controlling and converting electrical power is known as power electronics. There are primarily two areas of applications of power electronics such as switches or amplifiers. Ideally, switches should not dissipate any power while they are open or closed. The current in amplifiers varies continuously depending upon the controlled input. Some of the systems which are based on power electronics are AC/DC converters, DC/AC converters, DC/DC converters and AC/AC converters. Inverters are a type of devices which are used to convert DC to AC. This book discusses the fundamentals as well as modern approaches of power electronics. Those in search of information to further their knowledge will be greatly assisted by it. Coherent flow of topics, student-friendly language and extensive use of examples make this book an invaluable source of knowledge.

A foreword of all Chapters of the book is provided below:

Chapter 1 - The branch of electrical engineering which deals with the conversion of power from one form to the other is known as power electronics. This is an introductory chapter which will introduce briefly all the significant aspects of power electronics as well as their applications.; **Chapter 2** - Some of the fundamental concepts which are dealt with in power electronics are power, voltage, resistance and capacitance. The topics elaborated in this chapter will help in gaining a better perspective about the different concepts in power electronics as well as the different types of current.; **Chapter 3** - The semiconductor device which is used as a rectifier or a switch in power electronics is called a power semiconductor device. Some of the power semiconductor devices are diodes, transistors and thyristors. All the diverse principles of these power semiconductor devices have been carefully analyzed in this chapter.; **Chapter 4** - An electrical circuit which changes the electric energy from one form into the required form optimized for the specific load is known as a power converter. There are numerous types of power converters such as analog to digital converter, digital-to-analog converter and inverter. The topics elaborated in this chapter will help in gaining a better perspective about these types of power converters.; **Chapter 5** - The process through which usable electric current is distributed and produced is known as power generation. It can be generated through various sources such as solar energy, wind energy and geothermal energy. All the diverse principles of power generation and distribution have been carefully analyzed in this chapter.; **Chapter 6** - Electric power quality deals with the voltage, waveform and frequency of electricity. The process through which data on power quality is gathered, analyzed and interpreted is known as power quality monitoring. This chapter has been carefully written to provide an easy understanding of the varied facets of electric power quality and its control.; **Chapter 7** - Power electronics are applied in a variety of fields for converting AC current to DC, DC to AC, AC to AC and DC to DC. Some of the fields where power electronics are used are medical science, automotive and electrical engineering. The diverse applications of power electronics in these fields have been thoroughly discussed in this chapter.

I would like to thank the entire editorial team who made sincere efforts for this book and my family who supported me in my efforts of working on this book. I take this opportunity to thank all those who have been a guiding force throughout my life.

Giani Smith

Introduction to Power Electronics

The branch of electrical engineering which deals with the conversion of power from one form to the other is known as power electronics. This is an introductory chapter which will introduce briefly all the significant aspects of power electronics as well as their applications.

Power Electronics refers to the process of controlling the flow of current and voltage and converting it to a form that is suitable for user loads. The most desirable power electronic system is one whose efficiency and reliability is 100%.

The following block diagram shows the components of a Power Electronic system and how they are interlinked.

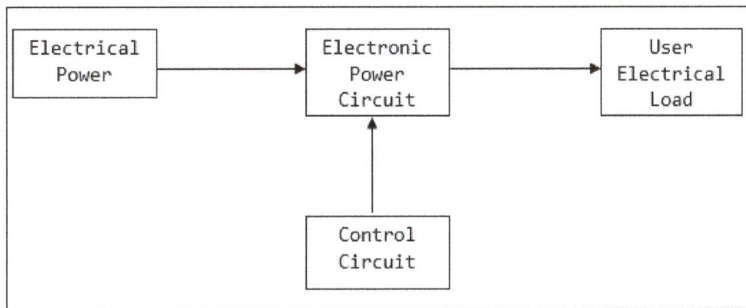

```
┌──────────────────────────────────────────────────────────────┐
│  ┌───────────┐      ┌───────────┐        ┌───────────┐        │
│  │ Electrical│      │ Electronic│        │   User    │        │
│  │   Power   │─────▶│   Power   │───────▶│ Electrical│        │
│  │           │      │  Circuit  │        │   Load    │        │
│  └───────────┘      └───────────┘        └───────────┘        │
│                          ▲                                    │
│                          │                                    │
│                     ┌───────────┐                             │
│                     │  Control  │                             │
│                     │  Circuit  │                             │
│                     └───────────┘                             │
└──────────────────────────────────────────────────────────────┘
```

A power electronic system converts electrical energy from one form to another and ensures the following is achieved:

- Maximum efficiency
- Maximum reliability
- Maximum availability
- Minimum cost
- Least weight
- Small size

Example:

A passenger lift in a modern building equipped with a Variable-Voltage-Variable-Speed induction-machine drive offers a comfortable ride and stops exactly at the floor level. Behind the scene it consumes less power with reduced stresses on the motor and corruption of the utility mains.

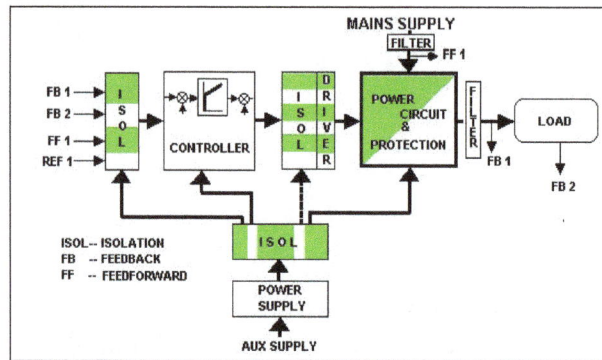

The block diagram of a typical Power Electronic converter

Power Electronics involves the study of:

- Power semiconductor devices - their physics, characteristics, drive requirements and their protection for optimum utilisation of their capacities,

- Power converter topologies involving them,

- Control strategies of the converters,

- Digital, analogue and microelectronics involved,

- Capacitive and magnetic energy storage elements,

- Rotating and static electrical devices,

- Quality of waveforms generated,

- Electro Magnetic and Radio Frequency Interference,

- Thermal Management.

The typical converter in figure illustrates the multidisciplinary nature of this subject.

Difference between Power Electronics and Linear Electronics

It is not primarily in their power handling capacities.

While power management IC's in mobile sets working on Power Electronic principles are meant to handle only a few milliwatts, large linear audio amplifiers are rated at a few thousand watts.

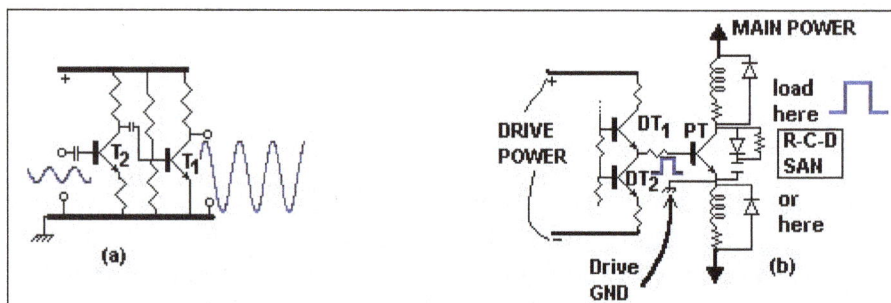

Typical Bipolar transistor based (a) linear (common emitter) (voltage) amplifier stage and
(b) switching (power) amplifier

The utilisation of the Bipolar junction transistor, figure in the two types of amplifiers best symbolises the difference. In Power Electronics all devices are operated in the switching mode - either 'FULLY-ON' or 'FULLY-OFF' states. The linear amplifier concentrates on fidelity in signal amplification, requiring transistors to operate strictly in the linear (active) zone, figure Saturation and cutoff zones in the V_{CE} - I_C plane are avoided. In a Power electronic switching amplifier, only those areas in the V_{CE} - I_C plane which have been skirted above, are suitable. Onstate dissipation is minimum if the device is in saturation (or quasi-saturation for optimising other losses). In the off-state also, losses are minimum if the BJT is reverse biased. A BJT switch will try to traverse the active zone as fast as possible to minimise switching losses.

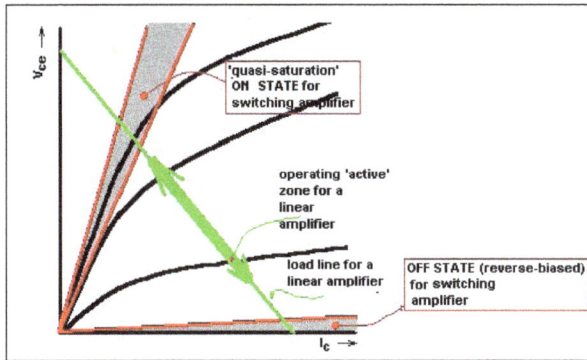

Operating zones for operating a Bipolar Junction Transistor as a linear and a switching amplifier

Linear operation	Switching operation
Active zone selected: Good linearity between input/output.	Active zone avoided: High losses, encountered only during transients.
Saturation & cut-off zones avoided: poor linearity.	Saturation & cut-off (negative bias) zones selected: low losses.
Transistor biased to operate around quiescent point.	No concept of quiescent point.
Common emitter, Common collector, common base modes.	Transistor driven directly at base – emitter and load either on collector or emitter.
Output transistor barely protected.	Switching-Aid-Network (SAN) and other protection to main transistor.
Utilisation of transistor rating of secondary importance.	Utilisation of transistor rating optimised.

An example illustrating the linear and switching solutions to a power supply specification will emphasise the difference.

Specification: Input: 230 V, 50 Hz and Output: 12 V regulated DC, 20 W.

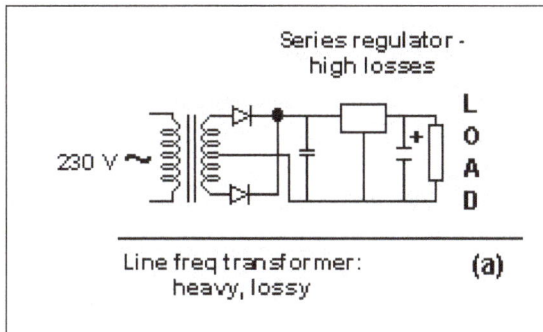

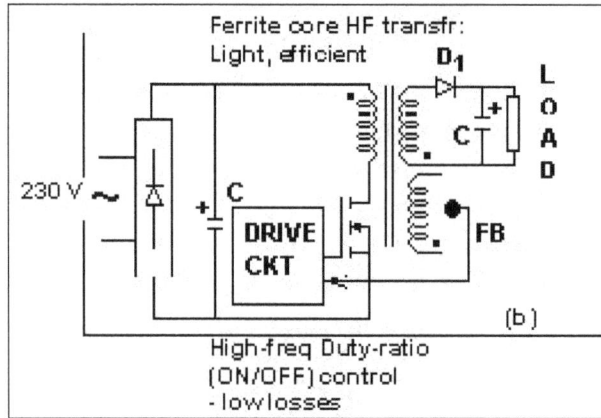

Ferrite core HF transfr:
Light, efficient

High-freq Duty-ratio
(ON/OFF) control
- low losses

(b)

(a) A Linear regulator and (b) a switching regulator solution of the specification above

The linear solution, figure (a), to this quite common specification would first step down the supply voltage to 12-0-12 V through a power frequency transformer. The output would be rectified using Power frequency diodes, electrolytic capacitor filter and then series regulated using a chip or a audio power transistor. The tantalum capacitor filter would follow. The balance of the voltage between the output of the rectifier and the output drops across the regulator device which also carries the full load current. The power loss is therefore considerable. Also, the step-down iron-core transformer is both heavy, and lossy. However, only twice-line-frequency ripples appear at the output and material cost and technical know-how required is low.

In the switching solution figure (b) using a MOSFET driven flyback converter, first the line voltage is rectified and then isolated, stepped-down and regulated. A ferrite-core high-frequency (HF) transformer is used. Losses are negligible compared to the first solution and the converter is extremely light. However significant high frequency (related to the switching frequency) noise appears at the output which can only be minimised through the use of costly 'grass' capacitors.

Applications of Power Electronics

- Static Applications: This utilizes moving and rotating mechanical parts such as welding, heating, cooling, and electro-plating and DC power.

- DC Power Supply:

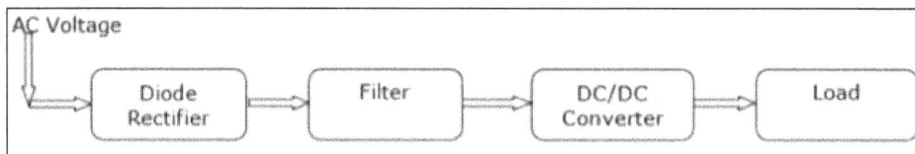

- Drive Applications: Drive applications have rotating parts such as motors. Examples include compressors, pumps, conveyer belts and air conditioning systems.

- Air Conditioning System: Power electronics is extensively used in air conditioners to control elements such as compressors. A schematic diagram that shows how power electronics is used in air conditioners is shown figure.

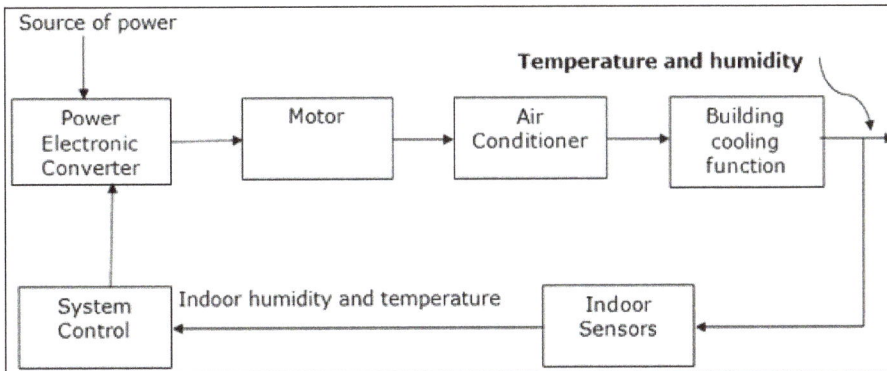

Growth Areas for Power Electronics

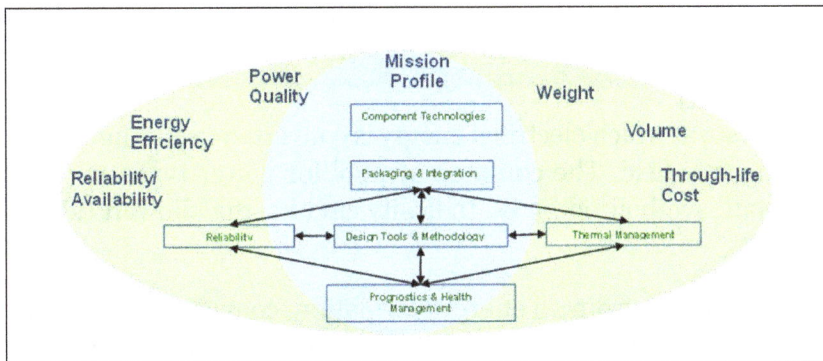

- Connection of renewable energy sources to power grids is not possible without power electronics.

- Future electricity networks must incorporate power electronics to maintain security of supply.

- Transport: electric and hybrid drive trains are only possible with efficient and intelligent power electronics. Weight savings through power electronics will reduce fuel demand.

- Power supplies: new concepts can improve overall efficiency by 2-4%.

- Motor drives: use 50-60% of all electrical energy consumed in the developed world: a potential reduction in energy consumption of 20- 30% is achievable.

- Lighting: power electronics can improve the efficiency of fluorescent, HID and LED ballasts by a minimum of 20%.

Basic Concepts of Power Electronics

Some of the fundamental concepts which are dealt with in power electronics are power, voltage, resistance and capacitance. The topics elaborated in this chapter will help in gaining a better perspective about the different concepts in power electronics as well as the different types of current.

Power

Electrical power is the rate at which electrical energy is converted to another form, such as motion, heat, or an electromagnetic field. The common symbol for power is the uppercase letter P. The standard unit is the watt, symbolized by W. In utility circuits, the kilowatt (kW) is often specified instead; 1 kW = 1000 W.

One watt is the power resulting from an energy dissipation, conversion, or storage process equivalent to one joule per second. When expressed in watts, power is sometimes called wattage. The wattage in a direct current (DC) circuit is equal to the product of the voltage in volts and the current in amperes. This rule also holds for low-frequency alternating current (AC) circuits in which energy is neither stored nor released. At high AC frequencies, in which energy is stored and released (as well as dissipated or converted), the expression for power is more complex.

In a DC circuit, a source of E volts, delivering I amperes, produces P watts according to the formula:

$$P = EI$$

When a current of I amperes passes through a resistance of Rohms, then the power in watts dissipated or converted by that component is given by:

$$P = I^2 R$$

When a potential difference of E volts appears across a component having a resistance of R ohms, then the power in watts dissipated or converted by that component is given by:

$$P = E^2 / R$$

In a DC circuit, power is a scalar (one-dimensional) quantity. In the general AC case, the determination of power requires two dimensions, because AC power is a vector quantity. Assuming there is no reactance (opposition to AC but not to DC) in an AC circuit, the power can be calculated according to the above formulas for DC, using root-mean-square values for the alternating current and voltage. If reactance exists, some power is alternately stored and released by the system. This is called apparent power or reactive power. The resistance dissipates power as heat or converts it to some other tangible form; this is called true power. The vector combination of reactance and resistance is known as impedance.

Voltage

Voltage is the pressure from an electrical circuit's power source that pushes charged electrons (current) through a conducting loop, enabling them to do work such as illuminating a light.

In brief, voltage = pressure, and it is measured in volts (V). The term recognizes Italian physicist Alessandro Volta (1745-1827), inventor of the voltaic pile—the forerunner of today's household battery.

In electricity's early days, voltage was known as electromotive force (emf). This is why in equations such as Ohm's Law; voltage is represented by the symbol E.

Example of voltage in a simple direct current (dc) circuit:

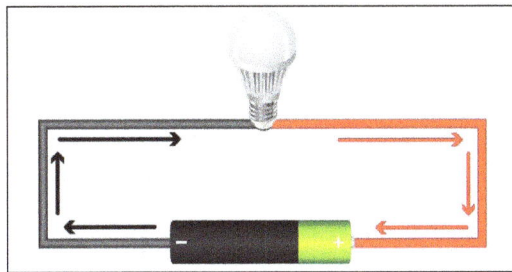

- In this dc circuit, the switch is closed (turned ON).

- Voltage in the power source—the "potential difference" between the battery's two poles—is activated, creating pressure that forces electrons to flow as current out the battery's negative terminal.

- Current reaches the light, causing it to glow.

- Current returns to the power source.

Voltage is either alternating current (ac) voltage or direct current (dc) voltage. Ways they differ:

Alternating current voltage (represented on a digital multimeter by $\widetilde{V}$):

- Flows in evenly undulating since waves, as shown below:

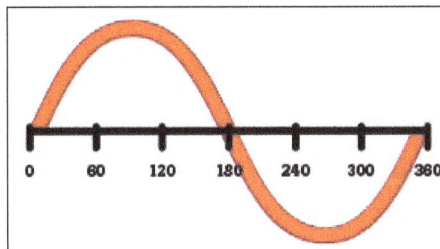

- Reverses direction at regular intervals.

- Commonly produced by utilities via generators, where mechanical energy—rotating motion powered by flowing water, steam, wind or heat—is converted to electrical energy.

- More common than dc voltage. Utilities deliver ac voltage to homes and businesses where the majority of devices use ac voltage.

- Primary voltage supplies vary by nation. In the United States, for example, it's 120 volts.

- Some household devices, such as TVs and computers, utilize dc voltage power. They use rectifiers (such as that chunky block in a laptop computer's cord) to convert ac voltage and current to dc.

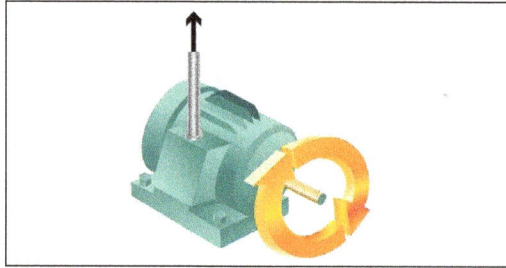

Generators convert rotating motion into electricity. The rotary motion is commonly caused by flowing water (hydroelectric power) or steam from water heated by gas, oil, coal or nuclear power.

Direct current voltage (represented on a digital multimeter by $\overline{\overline{V}}$ and $\overline{\overline{m}}V$):

- Travels in a straight line and in one direction only.

- Commonly produced by sources of stored energy such as batteries.

- Sources of dc voltage have positive and negative terminals. Terminals establish polarity in a circuit, and polarity can be used to determine if a circuit is dc or ac.

- Commonly used in battery-powered portable equipment (autos, flashlights, cameras).

Potential Difference

Voltage and the term "potential difference" are often used interchangeably. Potential difference might be better defined as the potential energy difference between two points in a circuit. The amount of difference (expressed in volts) determines how much potential energy exists to move electrons from one specific point to another. The quantity identifies how much work, potentially, can be done through the circuit.

A household AA alkaline battery, for example, offers 1.5 V. Typical household electrical outlets offer 120 V. The greater the voltage in a circuit, the greater its ability to "push" more electrons and do work.

Voltage/potential difference can be compared to water stored in a tank. The larger the tank, and the greater its height (and thus its potential velocity), the greater the water's capacity to create an impact when a valve is opened and the water (like electrons) can flow.

Need of Voltage Measurement

Technicians approach most troubleshooting situations knowing how a circuit should customarily perform.

Circuits are used to deliver energy to a load—from a small device to a household appliance to an industrial motor. Loads often carry a nameplate that identifies their standard electrical reference values, including voltage and current. In place of a nameplate, some manufacturers provide a detailed schematic (technical diagram) of a load's circuitry. Manuals may include standard values.

These numbers tell a technician what readings to expect when a load is operating normally. A reading on a digital multimeter can objectively identify deviations from the norm. Even so, the technician must use knowledge and experience to determine the factors causing such variances.

Volt

The volt (symbolized V) is the Standard International (SI) unit of electric potential or electromotive force. A potential of one volt appears across a resistance of one ohm when a current of one ampere flows through that resistance. Reduced to SI base units, $1V = 1kg$ times m^2 times s^{-3} times A^{-1} (kilogram meter squared per second cubed per ampere).

Voltage can be expressed as an average value over a given time interval, as an instantaneous value at a specific moment in time, or as an effective or root-mean-square (rms) value. Average and instantaneous voltages are assigned a polarity either negative (-) or positive (+) with respect to a zero, or ground, reference potential. The rms voltage is a dimensionless quantity, always represented by a non-negative real number.

For a steady source of direct-current (DC) electric potential, such as that from a zinc-carbon or alkaline electrochemical cell, the average and instantaneous voltages are both approximately +1.5 V if the negative terminal is considered the common ground; the rms voltage is 1.5 V. For standard utility alternating current (AC), the average voltage is zero (the polarity constantly reverses); the instantaneous voltage ranges between approximately -165 V and +165 V; the rms voltage is nominally 117 V.

Voltages are sometimes expressed in units representing power-of-10 multiples or fractions of one volt. A kilovolt (symbolized kV) is equal to one thousand volts ($1 kV = 10^3$ V). A megavolt (symbolized MV) is equal to one million volts ($1 MV = 10^6$ V). A millivolt (symbolized mV) is equal to one-thousandth of a volt ($1 mV = 10^{-3}$ V). A microvolt (symbolized μV) is equal to one-millionth of a volt ($1 μV = 10^{-6}$ V).

Electric Current

An electric current is a flow of electric charge in a circuit. More specifically, the electric current is the rate of charge flow past a given point in an electric circuit. The charge can be negatively charged electrons or positive charge carriers including protons, positive ions or holes.

The magnitude of the electric current is measured in coulombs per second, the common unit for this being the Ampere or amp which is designated by the letter 'A'.

The Ampere or amp is widely used within electrical and electronic technology along with the multipliers like milliamp (0.001A), microamp (0.000001A), and so forth.

Current flow in a circuit is normally designated by the letter 'I', and this letter is used in equations like Ohms law where $V=I·R$.

The basic concept of current is that it is the movement of electrons within a substance. Electrons are minute particles that exist as part of the molecular structure of materials. Sometimes these electrons are held tightly within the molecules and other times they are held loosely and they are able to move around the structure relatively freely.

Electrons are charged particles - they carry a negative charge. If they move then an amount of charge moves and this is called current.

It is also worth noting that the number of electrons that able to move governs the ability of a particular substance to conduct electricity. Some materials allow current to move better than others.

The motion of the free electrons is normally very haphazard - it is random - as many electrons move in one direction as in another and as a result there is no overall movement of charge.

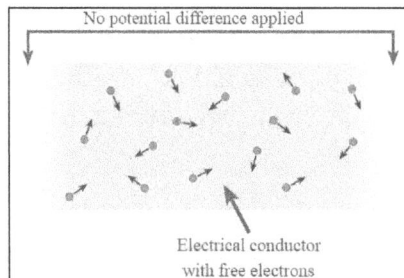

Random electron movement in a conductor with free electrons

If a force acts on the electrons to move them in a particular direction, then they will all drift in the same direction, although still in a somewhat haphazard fashion, but there is an overall movement in one direction.

The force that acts on the electrons is called and electromotive force, or EMF, and its quantity is voltage measured in volts.

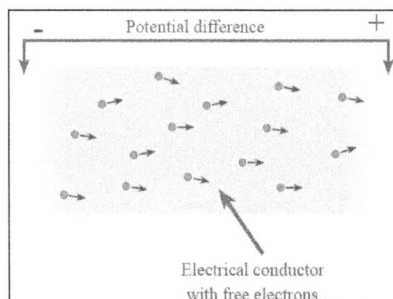

Electron flow under the action of applied electro-motive force

To gain a little more understanding about what current is and how it acts in a conductor, it can be compared to water flow in a pipe. There are limitations to this comparison, but it serves as a very basic illustration of current and current flow.

The current can be considered to be like water flowing through a pipe. When pressure is placed on one end it forces the water to move in one direction and flow through the pipe. The amount of water flow is proportional to the pressure placed on the end. The pressure or force placed on the end can be likened to the electro-motive force.

When the pressure is applied to the pipe, or the water is allowed to flow as a result of a tap being opened, then the water flows virtually instantaneously. The same is true for the electrical current.

To gain an idea of the flow of electrons, it takes 6.24 billion, billion electrons per second to flow for a current of one ampere.

Conventional Current and Electron Flow

The particles that carry charge along conductors are free electrons. The electric field direction within a circuit is by definition the direction that positive test charges are pushed. Thus, these negatively charged electrons move in the direction opposite the electric field.

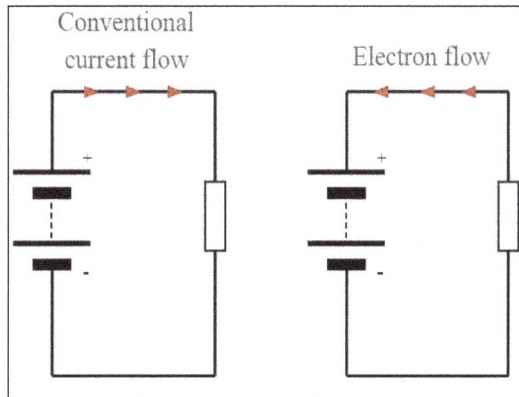

Electron and conventional current flow

This came about because the initial investigations in static and dynamic electric currents was based upon what we would now call positive charge carriers. This meant that then early convention for the direction of an electric current was established as the direction that positive charges would move.

- Conventional current flow: The conventional current flow is from positive to the negative terminal and indicates the direction that positive charges would flow.

- Electron flow: The electron flow is from negative to positive terminal. Electrons are negatively charged and are therefore attracted to the positive terminal as unlike charges attract.

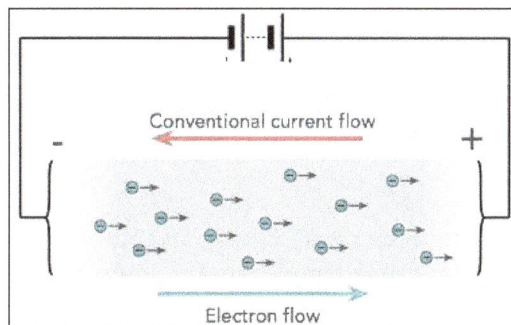

This is the convention that is used globally to this day, even though it may seem a little odd and out-dated.

Speed of Electron or Charge Movement

The speed of the transmission of electrical current is very different to that of the speed of the actual electron movement. The electron itself bounces around in the conductor, and possibly only makes progress along the conductor at the rate of a few millimetres a second. This means that in the case of alternating current, where the current changes direction 50 or 60 times per second, most of the electrons never make it out of the wire.

To take a different example, in the near-vacuum inside a cathode ray tube, the electrons travel in near-straight lines at about a tenth of the speed of light.

Effects of Current

When an electric current flows through a conductor there are a number of signs which tell that a current is flowing.

- Heat is dissipated: Possibly the most obvious is that heat is generated. If the current is small then the amount of heat generated is likely to be very small and may not be noticed. However if the current is larger then it is possible that a noticeable amount of heat is generated. An electric fire is a prime example showing how a current causes heat to be generated. The actual amount of heat is governed not only be the current, but also be the voltage and the resistance of the conductor.

- Magnetic effect: Another effect which can be noticed is that a magnetic field is built up around the conductor. If a current is flowing in conductor then it is possible to detect this. By placing a compass close to a wire carrying a reasonably large direct current, the compass needle can be seen to be deflect. Too fast for the needle to respond and the two wires (live and neutral) close together in the same cable will cancel out the field.

The magnetic field generated by a current is put to good use in a number of areas. By winding a wire into a coil, the effect can be increased, and an electro-magnet can be made. Relays and a host of other items use the effect. Loudspeakers also use a varying current in a coil to cause vibrations to occur in a diaphragm which enable the electronic currents to be converted into sounds.

Alternating Current

The current, which reverses its direction periodically in second, is called "Alternating Current". The polarity of the voltage is reversed periodically in an Alternating Current source. So the current reverses its direction periodically. Hence the name Alternating Current (AC) indicating current is alternating (periodically).

Number of times this current changes its direction in one second can be defined as the frequency of the alternating current. 50Hz frequency means it changes it direction for 50 times in a second.

In this current, the charged particles move from zero to maximum value in a direction and fall to zero and goes for another cycle in the opposite direction with the same values (in negative) . The values processed by AC current in both the directions are equal at all instances.

It has both magnitude, direction and it varies in accordance with time. AC will flow in both positive and negative directions and hence it is Bi-directional. Generally AC waveform is represented by Sinusoidal waveform and its mathematical formula is:

$$A(t) = A \sin (2\pi f t)$$

Where,

A is magnitude of signal,

t is the time period,

f is the frequency of signal.

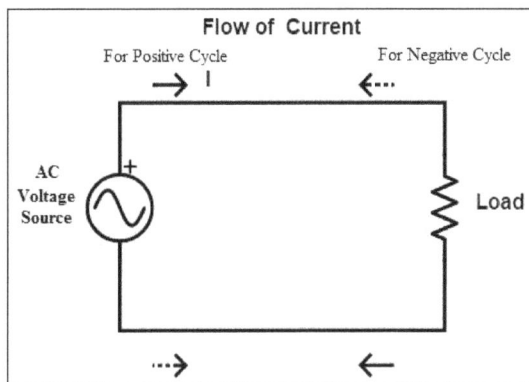

Generation of Alternating Current

Alternating current generating devices are called Alternators. The alternating current can be generated by many methods and by using many circuits. The simplest method of generating AC current is 'Using Basic Single Coil AC Generator'.

Working Principle

An AC generator works depending upon the principle of Faraday's electromagnetic induction to convert mechanical energy (such as rotation) into electrical energy.

Construction

A simple generator consists of two permanent magnets. They will produce a fixed magnetic field between north & south poles. In this magnetic field, there will be a single rectangular loop of wire which is an important part of the alternator.

It will rotate around a fixed axis in the magnetic field generated by the pole magnets and it cuts the magnetic flux. The central axis is perpendicular to the magnetic field generated by the two poles.

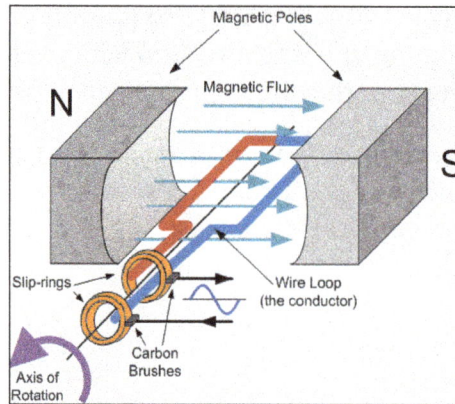

Process of Alternating Current Generation

- In the beginning, the wire is parallel to the field and does not cut magnetic flux i.e. the lines of magnetic force. So there is no voltage induced in the loop.

- As the wire rotates in a counter-clockwise direction, the wire sides will cut the magnetic lines of force in opposite directions.

- The induced EMF in the loop at any instant of time is proportional to the angle of rotation of the wire loop.

- When the wire loop has rotated for 180° across the magnetic lines of force in the opposite direction, then electrons in the wire flow in the opposite direction.

- This means, for every complete revolution (360°) of wire, one full sinusoidal waveform is produced.

- The rotating wire in magnetic field always connected with the carbon brushes and slip rings in all times.

- These are used to transfer the electrical current induced in the coil.

- The amount of EMF induced in a coil is determined by its speed, strength, and length.

- We know that 'frequency' is Number of cycles per second.

- Number of cycles produced will depend upon the speed of the rotating wire. That means the more the speed of the coil; the more will be the frequency. That means f is directly proportional to n, where 'n' number of revolutions of wire.

- The periodic reversal of polarity results in the generation of a voltage that has alternating polarity and hence the alternating current.

AC Wave Form

The alternating current can be represented by a waveform, with its amplitude and time period. There are many wave forms with which we can represent the AC current like square wave or triangular wave, but the easiest is to represent it with a sine wave.

For every complete cycle of the sine wave, we see a positive half and a negative half cycle. These positive and negative half cycles will represent that the AC current will change its direction periodically.

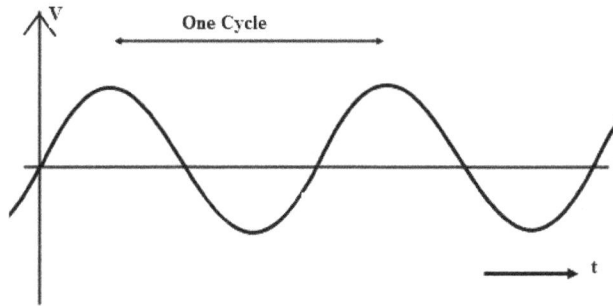

Advantages of AC over DC

- If we consider the cost of generating AC and DC currents, the cost of AC current generation is cheaper than that of DC current generation.

- Machines using Alternating current are simple in design and their maintenance is also easy compared to machines using DC current.

- We can use the AC current for long distance transmission, but not DC because the voltage range of alternating current is high.

- AC current can be easily transformed from low to high voltage or high to low voltage by using a transformer. Hence the range of voltages in AC current is very wide.

- Without any loss we can convert the AC current to DC by using rectifiers and we can limit the magnitude of AC current with the help of an inductor or conductor at a very considerable loss.

- Long distance transmission losses in AC current are very less compared to DC transmission, when the supplied voltage is high.

- AC motors will produce a higher output (higher torque) than DC motors.

Disadvantages

- The peak value of AC current is high, so it is very dangerous to use without insulation.

- AC transmission cost is very high compared to DC transmission.

- It a gives repelling shock of high voltage to a person who touches it and can cause electrocution.

- There is a chance of interference with communication signals.

Applications of AC current

- All the electrical appliances we use in our day to day life work with AC current supply only. Ex: T.V., refrigerator, bulb, fan, Air conditioner etc.

- Alternating current is used in equipments where high power signals are transmitted. We use the Alternating Current in high power distribution systems like Antenna wave propagation, radio waves and ultrasonic wave propagation.

- We also use AC current in transformers for long distance transmission, to supply for many other purposes.

- Mainly we use the type of current in industries where the power requirement is high.

Direct Current

DC stands for Direct Current, which is electrical current that flows in one direction.

In dc circuits, the current is in one direction unlike the alternating current (AC) where the current reverses direction 50 or 60 times a second depending on the frequency of the supply. As the direct current flows, the electrons, which constitute the electric charge, flow from the point of low potential to the point of high potential. They move from the negative terminal to the positive terminal and the resulting current is in the opposite direction (from positive to negative).

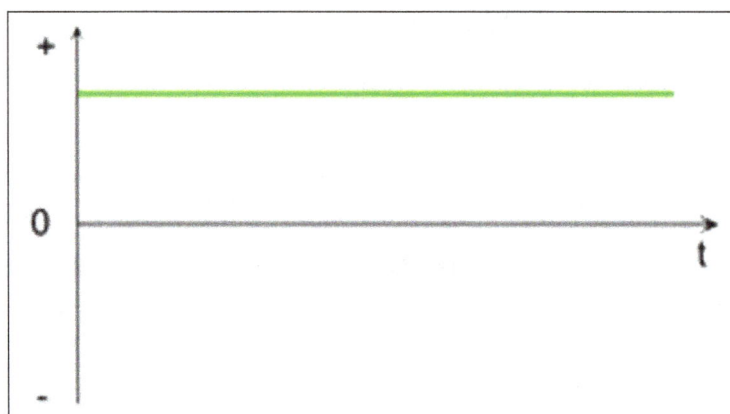

Plot of direct current/voltage against time

The DC is usually used in low-voltage applications such as battery operated equipment. When a DC is required, and the use of battery is uneconomical or requires constant charging, a power supply is used to convert the ac to dc. This will charge the battery, and the equipment used off the mains until the battery is depleted (i.e. a mobile phone). The other option is to power the circuit directly from the rectified DC and have the battery work as a backup power when the AC mains supply is unavailable (i.e. in a laptop).

Power supplies range from simple regulators with one output, to smps power supplies that are more regulated and capable of supplying multiple outputs such as those required to power a computer. The voltages and current's capacity of a power supply depend on the design and components used. There are a variety of DC voltage levels, typical values are 1.2, 1.5, 3, 3.3, 3.6, 5, 6, 10, 12, 15, 18, 18.5, 19, 20, 24, and 48 Volts.

An ideal DC supply should provide a constant voltage and adequate current throughout the operation of the equipment. However, sources such as batteries have a limited capacity and can only power the equipment efficiently for a given period determined by the rating of the battery and the load.

To maintain the power at a constant level, most equipment uses rechargeable batteries so that the charge can be replenished regularly. The charger consists of a rectifier circuit which converts the readily available AC into a suitable DC voltage.

Other than changing the mains AC to DC, most power supplies change the level of the voltage. Most of them will always lower the voltage since most of the electronics functions from much less voltage levels, but higher currents. Most laptops use between 18 Volts DC to 20 V DC and current of at least 3A. The laptop adapter must therefore provide a means of lowering the mains voltage from either the 220V or 120 Volts down to about 20 Volts DC.

Sources of Direct Current

- DC generators

- Batteries

- DC power converters that rectify the AC

- Solar panels

- Thermocouples.

Advantages of DC

- Most equipment are more efficient when directly powered from DC.

- More efficient especially when the power leads are less than 100 feet.

- Less risks of shock hazard for voltages below 48 Volts.

Disadvantages of DC

- High cost when dealing with high current systems; i.e. larger cables, fuses, switches and other components required for high current, low voltage applications.

- Difficult to obtain DC appliances and equipment.

- Higher chances of fire hazards when maximum circuit ratings and wire sizes are not strictly adhered to.

Applications

DC is used in almost all electronics equipment, electric vehicles, automation, control of electrical equipment, and more. Most office and domestic equipment such as TVs, audio systems, amplifiers, flash lights, computers, tablets and smartphones use DC power to function. However, since the universally available power is AC, the equipment uses an external or internal power supply to convert the utility mains AC into the desired DC for the equipment.

Comparision of AC and DC

	Alternating Current	Direct Current
Amount of energy that can be carried	Safe to transfer over longer city distances and can provide more power.	Voltage of DC cannot travel very far until it begins to lose energy.
Cause of the direction of flow of electrons	Rotating magnet along the wire.	Steady magnetism along the wire.
Frequency	The frequency of alternating current is 50Hz or 60Hz depending upon the country.	The frequency of direct current is zero.
Direction	It reverses its direction while flowing in a circuit.	It flows in one direction in the circuit.
Current	It is the current of magnitude varying with time.	It is the current of constant magnitude.
Flow of Electrons	Electrons keep switching directions - forward and backward.	Electrons move steadily in one direction or 'forward'.
Obtained from	A.C Generator and mains.	Cell or Battery.
Passive Parameters	Impedance.	Resistance only.
Power Factor	Lies between 0 & 1.	it is always 1.
Types	Sinusoidal, Trapezoidal, Triangular, Square.	Pure and pulsating.

Origins of AC and DC Current

A magnetic field near a wire causes electrons to flow in a single direction along the wire, because they are repelled by the negative side of a magnet and attracted toward the positive side. This is how DC power from a battery was born, primarily attributed to Thomas Edison's work.

AC generators gradually replaced Edison's DC battery system because AC is safer to transfer over the longer city distances and can provide more power. Instead of applying the magnetism along the wire steadily, scientist Nikola Tesla used a rotating magnet. When the magnet was oriented in one direction, the electrons flowed towards the positive, but when the magnet's orientation was flipped, the electrons turned as well.

Use of Transformers with Alternating Current

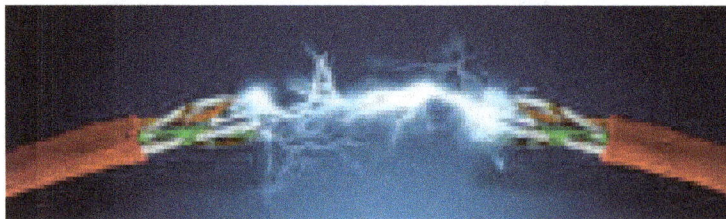

Another difference between AC and DC involves the amount of energy it can carry. Each battery is designed to produce only one voltage, and that voltage of DC cannot travel very far until it begins to lose energy. But AC's voltage from a generator, in a power plant, can be bumped up or down in strength by another mechanism called a transformer. Transformers are located on the electrical pole on the street, not at the power plant. They change very high voltage into a lower voltage appropriate for your home appliances, like lamps and refrigerators.

Storage and Conversion from AC to DC and Vice Versa

AC can even be changed to DC by an adapter that you might use to power the battery on your laptop. DC can be "bumped" up or down, it is just a little more difficult. Inverters change DC to AC. For example, for your car an inverter would change the 12 volt DC to 120 Volt AC to run a small device. While DC can be stored in batteries, you cannot store AC.

Resistance

Resistance is the property of an electric circuit or part of a circuit that transforms electric energy into heat energy in opposing electric current. Resistance involves collisions of the current-carrying charged particles with fixed particles that make up the structure of the conductors. Resistance is often considered as localized in such devices as lamps, heaters, and resistors, in which it predominates, although it is characteristic of every part of a circuit, including connecting wires and electric transmission lines.

The dissipation of electric energy in the form of heat, even though small, affects the amount of electromotive force, or driving voltage, required to produce a given current through the circuit. In fact, the electromotive force V (measured in volts) across a circuit divided by the current I (amperes) through that circuit defines quantitatively the amount of electrical resistance R. Precisely, $R = V/I$. Thus, if a 12-volt battery steadily drives a two-ampere current through a length of wire, the wire has a resistance of six volts per ampere, or six ohms. The ohm is the common unit of electrical resistance, equivalent to one volt per ampere and represented by the capital Greek letter omega, Ω. The resistance of a wire is directly proportional to its length and inversely proportional to its cross-sectional area. Resistance also depends on the material of the conductor.

The resistance of a conductor, or circuit element, generally increases with increasing temperature. When cooled to extremely low temperatures, some conductors have zero resistance. Currents continue to flow in these substances, called superconductors, after removal of the applied electromotive force.

The reciprocal of the resistance, $1/R$, is called the conductance and is expressed in units of reciprocal ohm, called mho.

Laws of Resistance

The resistance of any substance depends on the following factors:

1. Length of the substance.

2. Cross sectional area of the substance.

3. The nature of material of the substance.

4. Temperature of the substance.

There are mainly four laws of resistance from which the resistivity or specific resistance of any substance can easily be determined.

First Law of Resistivity

The resistance of a substance is directly proportional to the length of the substance. electrical resistance R of a substance is:

$$R \propto L$$

Where, L is the length of the substance.

If the length of a substance is increased, the path travelled by the electrons is also increased. If electrons travel long, they collide more and consequently the number of electrons passing through the substance becomes less; hence current through the substance is reduced. In other words, the resistance of the substance increases with increasing length of the substance. This relation is also linear.

Second Law of Resistivity

The resistance of a substance is inversely proportional to the cross-sectional area of the substance. Electrical resistance R of a substance is,

$$R \propto \frac{1}{A}$$

Where, A is the cross-sectional area of the substance.

The current through any substance depends on the numbers of electrons pass through a cross-section of substance per unit time. So, if the cross section of any substance is larger then more electrons can cross the cross section. Passing of more electrons through a cross-section per unit time causes more current through the substance. For fixed voltage, more current means less electrical resistance and this relation is linear.

Combining these two laws we get:

$$R \propto \frac{L}{A} \Rightarrow R = \rho \frac{L}{A}$$

Where, ρ (rho) is the proportionality constant and known as resistivity or specific resistance of the material of the conductor or substance. Now if we put, L = 1 and A = 1 in the equation, we get, R = ρ. That means resistance of a material of unit length having unit cross-sectional area is equal to its resistivity or specific resistance. Resistivity of a material can alternatively be defined as the electrical resistance between opposite faces of a cube of unit volume of that material.

Third Law of Resistivity

The resistance of a substance is directly proportional to the resistivity of the materials by which the substance is made. The resistivity of all materials is not the same. It depends on the number of free electrons, and size of the atoms of the materials, types of bonding in the materials and many other factors of the material structures. If the resistivity of a material is high, the resistance offered by the substance made by this material is high and vice versa. This relation is also linear.

$$R \propto \rho$$

Fourth Law of Resistivity

The temperature of the substance also affects the resistance offered by the substance. This is because, the heat energy causes more inter-atomic vibration in the metal, and hence electrons get more obstruction during drifting from lower potential end to higher potential end. Hence, in metallic substance, resistance increases with increasing temperature. If the substance in non-metallic, with increasing temperature, the more covalent bonds are broken, these cause more free electrons in the material. Hence, resistance is decreased with increase in temperature.

That is why mentioning resistance of any substance without mentioning its temperature is meaningless.

Unit of Resistivity

The unit of resistivity can be easily determined form its equation:

$$R = \rho \frac{L}{A} \Rightarrow \rho = \frac{RA}{L}$$

In SI System of Unit:

$$\rho = \frac{R\,\Omega \times A\,m^2}{L\,m}$$

$$\Rightarrow \rho = \frac{RA}{L} \frac{\Omega - m^2}{m} \ or\ \Omega - m$$

The unit of resistivity is $\Omega - m$ in MKS system and $\Omega - cm$ in CGS system and $1\,\Omega - m = 100\,\Omega - cm$.

Table: List of Resistivity of Different Commonly used Materials.

Materials	Resistivity in $\mu\,\Omega$ – cm at 20 °C
Aluminium	2.82
Brass	6 to 8
Carbon	3k to 7k
Constantan	49
Copper	1.72
Gold	2.44
Iron	12.0
Lead	22.0
Manganin	42 to 74
Mercury	96
Nickel	7.8
Silver	1.6
Tungsten	5.51
Zinc	6.3

Capacitance

Capacitance is defined as the capability of an element to store electric charge. A capacitor store electric energy in the form of the electric field by the two electrodes of a capacitor, one as positive and other as negative. In other words, Capacitance is a measure of charge per unit voltage that can be stored in an element. It is denoted by (C), and its unit is Farad (F).

The capacitance is mainly classified into two types; they are the self-capacitance and the mutual capacitance. The substance which has more self-capacitance store more electric charges and substance which have low capacitance stores less electrical charges.

Explanation and Derivation of Capacitance

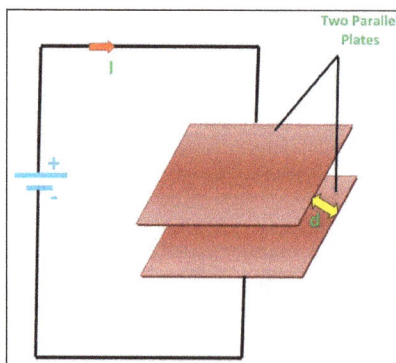

If the two parallel plates are connected, overlapping each other and are connected to a DC supply voltage as shown in the figure. The two plates are separated by an insulating dielectric so that the charge does not cross each other. One terminal of the parallel plate is connected to the positive supply and other to the negative supply. As the supply is ON, the capacitor starts charging, and it stores energy even if the supply is OFF.

The capacitance equation is given by:

$$C = \varepsilon_0 \varepsilon_r \frac{A}{d}$$

Where,

- C is the capacitance in Farad or Micro Farad.

- A is the overlapping area of the two plates in meter square.

- d is the distance of separation between two plates in meters.

- ε_0 is known as the electric constant.

- ε_r is the dielectric constant of the material between the two plates.

The capacitance is said to be one farad if one coulomb of charge is stored with one volt across the two electrodes of the element. The element which has capacitance is called a Capacitor.

The Charge on the capacitor at any instant of time is:

$$q = Cv$$

q being the amount of charge that can be stored in a capacitor of capacitance (C) against a potential difference of (v) volts.

$$i = C\frac{dv}{dt} \quad \left[as\, i = \frac{dq}{dt} \right]$$

Where, i, q and v represents the instantaneous value of current, charge and voltage respectively.

$$dv = \frac{1}{C}i\, dt \ \ \text{or} \ \int_{v_0}^{v_t} dv = \frac{1}{C}\int_0^t i\, dt$$

Where,

v_0 is the initial voltage of the capacitor.

v_t is the final voltage of the capacitor.

Now,

$$v_t - v_0 = \frac{1}{C}\int_0^t i\, dt$$

$$v_t = \frac{1}{C}\int_0^t i\, dt + v_0$$

The power absorbed by the capacitor is given by the equation shown below:

$$p = vi = v\,C\frac{dv}{dt}$$

The energy stored by the capacitor is given as:

$$W = \int_0^t P\,dt = \int_0^t v\,C\,\frac{dv}{dt}dt = \frac{1}{2}\,Cv^2$$

The current through the capacitor is zero if the applied voltage across the capacitor is constant. This means that when the DC voltage is applied across the capacitor with no initial charge, the capacitor first acts as a short circuit but as soon as it gets fully charged, the capacitor starts behaving as an open circuit.

The capacitor only stores energy and never dissipates the energy in any form. It can store a finite amount of energy, even if the current through the capacitor is zero.

Types of Capacitor

The various types of capacitor are as follows:

- Paper Capacitor
- Air capacitor
- Plastic Capacitor
- Silver mica capacitor
- Ceramic capacitor
- Electrolytic capacitor
- Porcelain capacitor.

Series and Parallel Capacitance in a Circuit

Series Capacitor Circuit

If the number of capacitors, for example, C_1, C_2, C_3....... connected together in a series is called a series capacitor circuit. The current flowing in this type of circuit will be same across all the capacitor as they are connected in series. The series connection of the capacitor is shown in the figure.

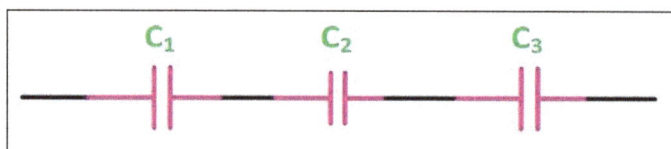

The equivalent capacitance is given by the equation as:

$$\frac{1}{C_{eq}} = \frac{1}{C_1} + \frac{1}{C_2} + \frac{1}{C_3} + \ldots \ldots \ldots \ldots$$

Parallel Capacitor Circuit

If the number of capacitors is connected to each other as in parallel connection, the circuit is said to be a Parallel Capacitor Circuit. The circuit is shown in the figure.

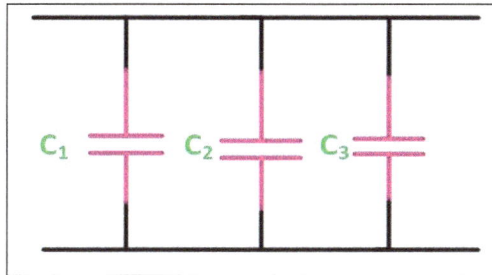

The equivalent capacitance in parallel circuit is given by the equation shown below:

$$C_{eq} = C_1 + C_2 + C_3 + \ldots \ldots \ldots$$

References

- Power: techtarget.com, Retrieved 2 April, 2019
- What-is-voltage, electricity, measurement-basics: fluke.com, Retrieved 12 June, 2019
- Volt: techtarget.com, Retrieved 22 February, 2019
- What-is-electrical-current, current: electronics-notes.com, Retrieved 3 March, 2019
- Introduction-to-alternating-current: electronicshub.org, Retrieved 13 July, 2019
- What-is-dc-direct-current: sunpower-uk.com, Retrieved 23 January, 2019
- Alternating-Current-vs-Direct-Current: diffen.com, Retrieved 4 June, 2019
- Resistance-electronics, technology: britannica.com, Retrieved 14 March, 2019
- Electrical-resistance-and-laws-of-resistance: electrical4u.com, Retrieved 24 August, 2019
- What-is-a-capacitance: circuitglobe.com, Retrieved 5 May, 2019

Power Semiconductor Devices

The semiconductor device which is used as a rectifier or a switch in power electronics is called a power semiconductor device. Some of the power semiconductor devices are diodes, transistors and thyristors. All the diverse principles of these power semiconductor devices have been carefully analyzed in this chapter.

Semiconductor Devices

The Semiconductor device is made up of a material that is neither a good conductor nor a good insulator, it is called a semiconductor. Such devices have established wide applications because of their reliability, compactness, and low cost. These are discrete components which are used in power devices, compactness optical sensors, and light emitters, including solid-state lasers. They have a wide range of current and voltage handling capabilities, with current ratings more than 5,000 amperes and voltage ratings more than 100,000 volts. More importantly, semiconductor devices lend themselves to integration into complex but readily build-up microelectronic circuits. They are having probable future, the key elements of the majority of electronic systems including communications with data-processing, consumer, and industrial-control equipment.

Semiconductor devices are electronic components that exploit the electronic properties of semi-conductor materials, like as silicon, germanium, and gallium arsenide, as well as organic semi-conductors. Semiconductor devices have replaced vacuum tubes in many applications. They use electronic conduction in the solid state as opposed to the thermionic emission in a high vacuum. Semiconductor devices are manufactured for both discrete devices and integrated circuits, which consist of from a few to billions of devices manufactured and interconnected on a single semiconductor substrate or wafer.

Semiconductor materials are useful by their behavior which can be easily manipulated by the addition of impurities is known as doping. Semiconductor conductivity can be controlled by the electric or magnetic field, by exposure to light or heat, or by the mechanical deformation of a doped mono crystalline grid; thus, semiconductors can make excellent sensors. Current conduction in a semiconductor occurs free of electrons and holes, collectively known as charge carriers. Doping of silicon is done by adding a small amount of impurity atoms and also for phosphorus or boron, significantly increases the number of electrons or holes within the semiconductor.

When a doped semiconductor contains excess holes it is called "p-type"(positive for holes)semiconductor, and when it contains some excess of free electrons, it is known as "n-type"(negative for electrons) semiconductor, is the sign of charge of the majority mobile charge carriers. The junctions which formed where n-type and p-type semiconductors are joined together is called p–n junction.

Semiconductor Device Materials

The silicon (Si) is most widely used material in semiconductor devices. It's having lower raw material cost and relatively simple process. Its useful temperature range makes it currently the best compromise among the various competing materials. Silicon used in semiconductor device manufacturing is presently fabricated into bowls that are large enough in diameter to allow the manufacture of 300 mm (12 in.) wafers.

Germanium (Ge) was a widely used in early semiconductor material, but its thermal sensitivity makes less useful than silicon. Nowadays, germanium is often alloyed with (Si) silicon for use in very-high-speed SiGe devices; IBM is a main producer of such devices.

Gallium arsenide (GaAs) is also widely used with high-speed devices, but so far, it has been difficult to form large-diameter bowls of this material, limiting the wafer diameter sizes significantly smaller than silicon wafers thus making mass production of Gallium arsenide (GaAs) devices significantly more expensive than silicon.

Most Common Semiconductor Devices

The list of common semiconductor devices mainly includes two terminals, three terminals and four terminal devices.

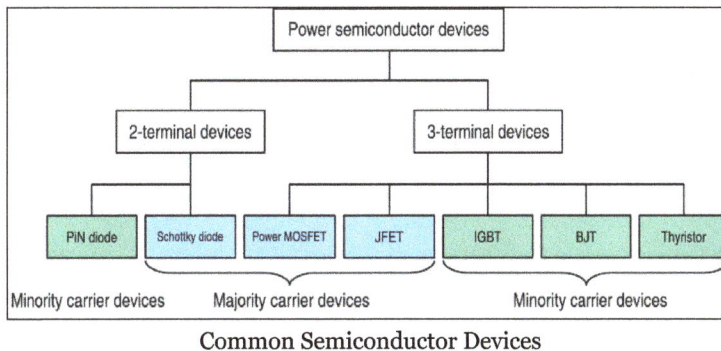

Common Semiconductor Devices

The two-terminal devices are:

- Diode (rectifier diode)
- Gunn diode
- IMPATT diode
- Laser diode
- Zener diode
- Schottky diode
- PIN diode
- Tunnel diode
- Light-emitting diode (LED)

- Photo transistor
- Photocell
- Solar cell
- Transient-voltage-suppression diode
- VCSEL.

Three-terminal devices are:

- Bipolar transistor
- Field-effect transistor
- Darlington transistor
- Insulated-gate bipolar transistor (IGBT)
- Unijunction transistor
- Silicon-controlled rectifier
- Thyristor
- TRIAC.

Four-terminal devices are:

- Photo coupler (Optocoupler)
- Hall effect sensor (magnetic field sensor).

Semiconductor Device Applications

All types of transistor can be used as the building blocks of logic gates, which is useful to design of digital circuits. In digital circuits like as microprocessors, transistors so which is acting as a switch (on-off); in the MOSFET, for example, the voltage applied to the gate determines whether the switch is on or off.

The transistors are used for analog circuits do not act as switches (on-off); relatively, they respond to a continuous range of input with a continuous range of output. Common analog circuits include oscillators and amplifiers. The circuits that interface or translate between analog circuits and digital circuits are known as the mixed-signal circuits.

Advantages of Semiconductor Devices

- As semiconductor devices have no filaments, hence no power is needed to heat them to cause the emission of electrons.
- Since no heating is required, semiconductor devices are set into operation as soon as the circuit is switched on.
- During operation, semiconductor devices do not produce any humming noise.

- Semiconductor devices require low voltage operation as compared to vacuum tubes.

- Owing to their small sizes, the circuits involving semiconductor devices are very compact.

- Semiconductor devices are shock proof.

- Semiconductor devices are cheaper as compared to vacuum tubes.

- Semiconductor devices have an almost unlimited life.

- As no vacuum has to be created in semiconductor devices, they have no vacuum deterioration trouble.

Disadvantages of Semiconductor Devices

- The noise level is higher in semiconductor devices as compared to that in the vacuum tubes.

- Ordinary semiconductor devices cannot handle as more power as ordinary vacuum tubes can do.

- In high frequency range, they have poor responder.

Diode

A diode is defined as a two-terminal electronic component that only conducts current in one direction (so long as it is operated within a specified voltage level). An ideal diode will have zero resistance in one direction, and infinite resistance in the reverse direction.

Although in the real world, diode's cannot achieve zero or infinite resistance. Instead, a diode will have negligible resistance in one direction (to allow current flow), and a very high resistance in the reverse direction (to prevent current flow). A diode is effectively like a valve for an electrical circuit.

Semiconductor diodes are the most common type of diode. These diodes begin conducting electricity only if a certain threshold voltage is present in the forward direction (i.e. the "low resistance" direction). The diode is said to be "forward biased" when conducting current in this direction. When connected within a circuit in the reverse direction (i.e. the "high resistance" direction), the diode is said to be "reverse biased".

A diode only blocks current in the reverse direction (i.e. when it is reverse biased) while the reverse voltage is within a specified range. Above this range, the reverse barrier breaks. The voltage at which this breakdown occurs is called the "reverse breakdown voltage". When the voltage of the circuit is higher than the reverse breakdown voltage, the diode is able to conduct electricity in the reverse direction (i.e. the "high resistance" direction). This is why in practice we say diode's have a high resistance in the reverse direction – not an infinite resistance.

A PN junction is the simplest form of the semiconductor diode. In ideal conditions, this PN junction behaves as a short circuit when it is forward biased and as an open circuit when it is in the reverse biased. The name diode is derived from "di–ode" which means a device that has two electrodes.

Diode Symbol

The symbol of a diode is shown below. The arrowhead points in the direction of conventional current flow in the forward biased condition. That means the anode is connected to the p side and cathode is connected to the n side.

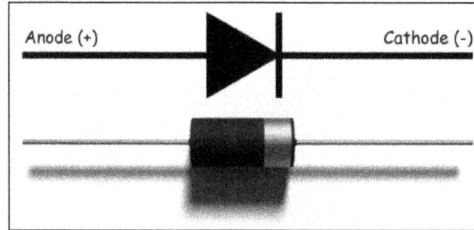

We can create a simple PN junction diode by doping pentavalent or donor impurity in one portion and trivalent or acceptor impurity in other portion of silicon or germanium crystal block. These dopings make a PN junction at the middle part of the block. We can also form a PN junction by joining a p-type and n-type semiconductor together with a special fabrication technique. The terminal connected to the p-type is the anode. The terminal connected to the n-type side is the cathode.

Working Principle of Diode

A diode's working principle depends on the interaction of n-type and p-type semiconductors. An n-type semiconductor has plenty of free electrons and a very few numbers of holes. In other words, we can say that the concentration of free electrons is high and that of holes is very low in an n-type semiconductor. Free electrons in the n-type semiconductor are referred as majority charge carriers, and holes in the n-type semiconductor are referred to as minority charge carriers.

A p-type semiconductor has a high concentration of holes and low concentration of free electrons. Holes in the p-type semiconductor are majority charge carriers, and free electrons in the p-type semiconductor are minority charge carriers.

Unbiased Diode

Now let us see what happens when one n-type region and one p-type region come in contact. Here due to concentration difference, majority carriers diffuse from one side to another. As the concentration of holes is high in the p-type region and it is low in the n-type region, the holes start diffusing from the p-type region to n-type region. Again the concentration of free electrons is high in the n-type region and it is low in the p-type region and due to this reason, free electrons start diffusing from the n-type region to the p-type region.

The free electrons diffusing into the p-type region from the n-type region would recombine with holes available there and create uncovered negative ions in the p-type region. In the same way, the holes diffusing into the n-type region from the p-type region would recombine with free electrons available there and create uncovered positive ions in the n-type region.

In this way, there would a layer of negative ions in the p-type side and a layer of positive ions in the n-type region appear along the junction line of these two types of semiconductor. The layers of uncovered positive ions and uncovered negative ions form a region at the middle of the diode where no charge carrier exists since all the charge carriers get recombined here in this region. Due to lack of charge carriers, this region is called depletion region.

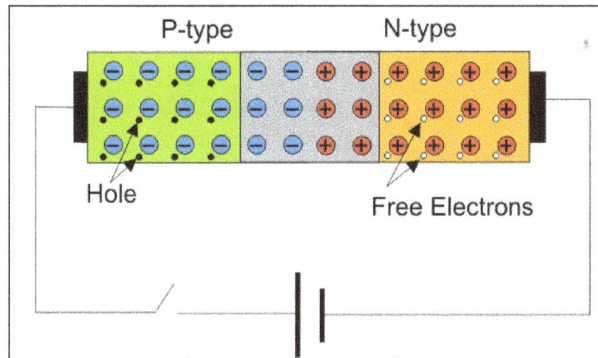

After the formation of the depletion region, there is no more diffusion of charge carriers from one side to another in the diode. This is because due to the electric field appeared across the depletion region will prevent further migration of charge carriers from one side to another. The potential of the layer of uncovered positive ions in the n-type side would repeal the holes in the p-type side and the potential of the layer of uncovered negative ions in the p-type side would repeal the free electrons in the n-type side. That means a potential barrier is created across the junction to prevent further diffusion of charge carriers.

Forward Biased Diode

Now let us see what happens if positive terminal of a source is connected to the p-type side and the negative terminal of the source is connected to the n-type side of the diode and if we increase the voltage of this source slowly from zero.

In the beginning, there is no current flowing through the diode. This is because although there is an external electrical field applied across the diode but still the majority charge carriers do not get sufficient influence of the external field to cross the depletion region. As we told that the depletion region acts as a potential barrier against the majority charge carriers. This potential barrier is called forward potential barrier. The majority charge carriers start crossing the forward potential barrier only when the value of externally applied voltage across the junction is more than the potential of the forward barrier. For silicon diodes, the forward barrier potential is 0.7 volt and for germanium diodes, it is 0.3 volt. When the externally applied forward voltage across the diode becomes more than the forward barrier potential, the free majority charge carriers start crossing the barrier and contribute the forward diode current. In that situation, the diode would behave as a short-circuited path and the forward current gets limited by only externally connected resistors to the diode.

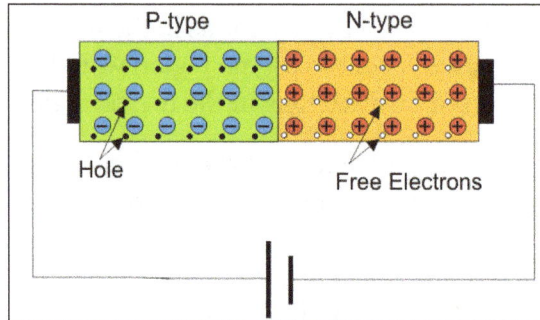

Reverse Biased Diode

Now let us see what happens if we connect negative terminal of the voltage source to the p-type side and positive terminal of the voltage source to the n-type side of the diode. At that condition, due to electrostatic attraction of negative potential of the source, the holes in the p-type region would be shifted more away from the junction leaving more uncovered negative ions at the junction. In the same way, the free electrons in the n-type region would be shifted more away from the junction towards the positive terminal of the voltage source leaving more uncovered positive ions in the junction. As a result of this phenomenon, the depletion region becomes wider. This condition of a diode is called the reverse biased condition. At that condition, no majority carriers cross the junction as they go away from the junction. In this way, a diode blocks the flow of current when it is reverse biased.

There are always some free electrons in the p-type semiconductor and some holes in the n-type semiconductor. These opposite charge carriers in a semiconductor are called minority charge carriers. In the reverse biased condition, the holes find themselves in the n-type side would easily cross the reverse biased depletion region as the field across the depletion region does not present rather it helps minority charge carriers to cross the depletion region. As a result, there is a tiny current flowing through the diode from positive to the negative side. The amplitude of this current is very small as the number of minority charge carriers in the diode is very small. This current is called reverse saturation current.

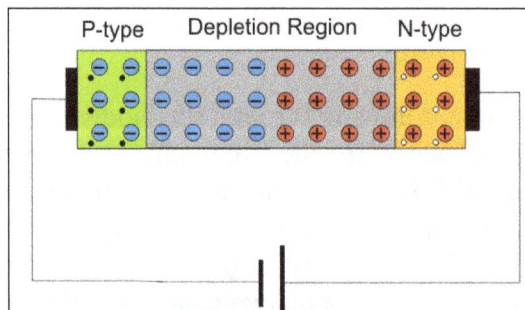

If the reverse voltage across a diode gets increased beyond a safe value, due to higher electrostatic force and due to a higher kinetic energy of minority charge carriers colliding with atoms, a number of covalent bonds get broken to contribute a huge number of free electron-hole pairs in the diode and the process is cumulative. The huge number of such generated charge carriers would contribute a huge reverse current in the diode. If this current is not limited by an external resistance connected to the diode circuit, the diode may permanently be destroyed.

Types of Diode

Zener Diode

Zener diode is basically like an ordinary PN junction diode but normally operated in reverse biased condition. But ordinary PN junction diode connected in reverse biased condition is not used as Zener diode practically. A Zener diode is a specially designed, highly doped PN junction diode.

Working Principle of Zener Diode

When a PN junction diode is reverse biased, the depletion layer becomes wider. If this reverse biased voltage across the diode is increased continually, the depletion layer becomes more and more wider. At the same time, there will be a constant reverse saturation current due to minority carriers.

After certain reverse voltage across the junction, the minority carriers get sufficient kinetic energy due to the strong electric field. Free electrons with sufficient kinetic energy collide with stationary ions of the depletion layer and knock out more free electrons. These newly created free electrons also get sufficient kinetic energy due to the same electric field, and they create more free electrons by collision cumulatively. Due to this commutative phenomenon, very soon, huge free electrons get created in the depletion layer, and the entire diode will become conductive. This type of breakdown of the depletion layer is known as avalanche breakdown, but this breakdown is not quite sharp. There is another type of breakdown in depletion layer which is sharper compared to avalanche breakdown, and this is called Zener breakdown. When a PN junction is diode is highly doped, the concentration of impurity atoms will be high in the crystal. This higher concentration of impurity atoms causes the higher concentration of ions in the depletion layer hence for same applied reverse biased voltage, the width of the depletion layer becomes thinner than that in a normally doped diode.

Due to this thinner depletion layer, voltage gradient or electric field strength across the depletion layer is quite high. If the reverse voltage is continued to increase, after a certain applied voltage, the electrons from the covalent bonds within the depletion region come out and make the depletion region conductive. This breakdown is called Zener breakdown. The voltage at which this breakdown occurs is called Zener voltage. If the applied reverse voltage across the diode is more than Zener voltage, the diode provides a conductive path to the current through it hence; there is no chance of further avalanche breakdown in it. Theoretically, Zener breakdown occurs at a lower voltage level then avalanche breakdown in a diode, especially doped for Zener breakdown. The Zener breakdown is much sharper than avalanche breakdown. The Zener voltage of the diode gets adjusted during manufacturing with the help of required and proper doping. When a zener diode is connected across a voltage source, and the source voltage is more than Zener voltage, the voltage across a Zener diode remain fixed irrespective of the source voltage. Although at that condition current through the diode can be of any value depending on the load connected with the diode. That is why we use a Zener diode mainly for controlling voltage in different circuits.

Zener Diode Circuit

Zener Diode is nothing but a single diode connected in a reverse bias, we have already stated that. A diode connected in reverse bias position in a circuit is shown in the figure.

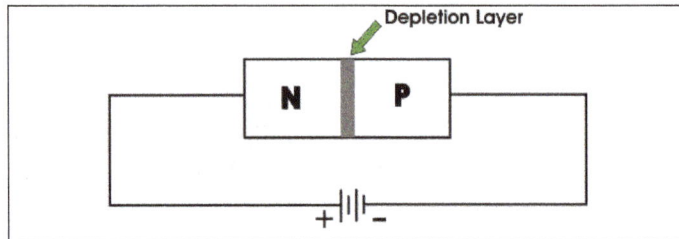

The circuit symbol of a Zener diode is also shown in the figure.

Characteristics of a Zener Diode

The diode circuits we should look through the graphical representation of the operation of the zener diode. Normally, it is called the V-I characteristics of a Zener diode.

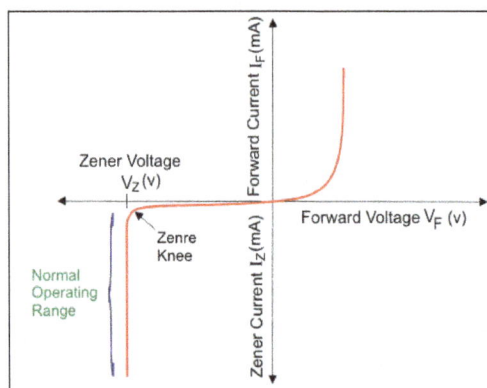

The diagram shows the V-I characteristics of a zener diode. When the diode is connected in forward bias, this diode acts as a normal diode but when the reverse bias voltage is greater than zener voltage, a sharp breakdown takes place. In the V-I characteristics above V_z is the zener voltage. It is also the knee voltage because at this point the current increases very rapidly.

P-N junction diode

PN junction diode is one of the fundamental elements in electronics. In this type of diode, we dope one side of a semiconductor piece with an acceptor impurity and another side with a donor impurity. A PN junction diode is a two-terminal electronics element, which can be classified as either 'step graded' or 'linear graded'.

In a step graded PN junction diode, the concentration of dopants both, in the N side and the P side are uniform up to the junction. But in a linearly graded junction, the doping concentration varies almost linearly with the distance from the junction. When we do not apply any voltage across the PN diode, free electrons will diffuse through the junction to P side and holes will diffuse through the junction to N side and they combine with each other.

Thus the acceptor atoms in the p-side near the junction edge and donor atoms in n-side near junction edge become negative and positive ions respectively. The existence of negative ions in the p-type side along the junction and positive ions in the n-type side along the junction edge creates an electric field. The electric filed opposes further diffusion of free electrons from the n-type side and holes from the p-type side of the PN junction diode. We call this region across the junction where the uncovered charges (ions) exist, as depletion region.

If, we apply forward bias voltage to the p-n junction diode. That means if the positive side of the battery is connected to the p-side, then the depletion regions width decreases and carriers (holes and free electrons) flow across the junction. If we apply a reverse bias voltage to the diode, the depletion width increases and no charge can flow across the junction.

P-N Junction Diode Characteristics

Let us consider a pn junction with a donor concentration N_D and acceptor concentration N_A. Let us also assume that all the donor atoms have donated free electrons and become positive donor ions and all the acceptor atoms have accepted electrons and created corresponding holes and become negative acceptor ions. So we can say the concentration of free electrons (n) and donor ions N_D are the same and similarly, the concentration of holes (p) and acceptor ions (N_A) are the same. Here, we have ignored the holes and free electrons created in the semiconductors due to unintentional impurities and defects.

$$n = N_D \ \ and \ \ p = N_A$$

Across the pn junction, the free electrons donated by donor atoms in n-type side diffuse to the p-typer side and recombine with holes. Similarly, the holes created by acceptor atoms in p-type side diffuse to the n-type side and recombine with free electrons. After this recombination process, there is a lack of or depletion of charge carriers (free electrons and holes) across the junction. The region across the junction where the free charge carriers get depleted is called depletion region. Due to the absence of free charge carriers (free electrons and holes), the donor ions of n-type side and acceptor ions of p-type side across the junction become uncovered. These positive uncovered donor ions towards n-type side adjacent to the junction and negative uncovered acceptors ions towards p-type side adjacent to the junction cause a space charge across the pn junction. The potential developed across the junction due to this space charge is called the diffusion voltage. The diffusion voltage across a pn junction diode can be expressed as:

$$V_D = \frac{kT}{e} \ln \frac{N_A N_D}{n_i^2}$$

The diffusion potential creates a potential barrier for further migration of free electrons from n-type side to p-type side and holes from p-type side to n-type side. That means diffusion potential prevents charge carriers to cross the junction. This region is highly resistive because of depletion of free charge carriers in this region. The width of the depletion region depends on the applied bias voltage. The relation between the width of the depletion region and bias voltage can be represented by an equation called Poisson Equation.

$$W_D = \sqrt{\frac{2\varepsilon}{e}(V_D - V)\left(\frac{1}{N_A} + \frac{1}{N_D}\right)}$$

Here, ε is the permittivity of the semiconductor and V is the biasing voltage. So, on an application of a forward bias voltage the width of the depletion region i.e. pn junction barrier decreases and ultimately disappears. Hence, in absence of potential barrier across the junction in the forward bias condition free electrons enter into the p-type region and holes enter into the n-type region, where they recombine and release a photon at each recombination. As a result, there will be a forward current flowing through the diode. The current through the PN junction is expressed as:

$$I = I_s \left(e^{\frac{eV}{kT}} - 1 \right)$$

Here, voltage V is applied across the pn junction and total current I, flows through the pn junction. I_s is reverse saturation current, e = charge of electron, k is Boltzmann constant and T is temperature in Kelvin scale.

The graph shows the current-voltage characteristic of a PN junction diode.

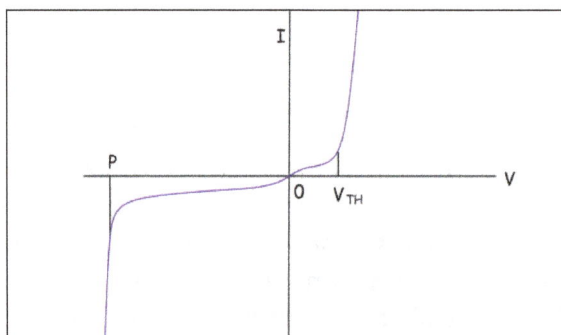

When, V is positive the junction is forward biased, and when V is negative, the junction is reverse biased. When V is negative and less than V_{TH}, the current is minimal. But when V exceeds V_{TH}, the current suddenly becomes very high. The voltage V_{TH} is known as the threshold or cut in voltage.

For Silicon diode V_{TH} = 0.6 V. At a reverse voltage corresponding to the point P, there is abrupt increment in reverse current. This portion of the characteristics is known as breakdown region.

Tunnel Diode

The application of transistors is very high in frequency range are hampered due to the transit time and other effects. Many devices use the negative conductance property of semiconductors for high frequency applications. Tunnel diode is one of the most commonly used negative conductance devices. It is also known as Esaki diode after L. Esaki for his work on this effect. This diode is a two terminal device. The concentration of dopants in both p and n region is very high. It is about 10^{24} – 10^{25} m^{-3} the pn junction is also abrupt. For this reasons, the depletion layer width is very small. In the current voltage characteristics of tunnel diode, we can find a negative slope region when forward bias is applied.

Quantum mechanical tunneling is responsible for the phenomenon and thus this device is named as tunnel diode. The doping is very high so at absolute zero temperature the Fermi levels lies within the bias of the semiconductors. When no bias is applied any current flows through the junction.

Characteristics of Tunnel Diode

When reverse bias is applied the Fermi level of p-side becomes higher than the Fermi level of n-side. Hence, the tunneling of electrons from the balance band of p-side to the conduction band of n-side takes place. With the interments of the reverse bias the tunnel current also increases. When forward junction is a applied the Fermi level of n-side becomes higher that the Fermi level of p-side thus the tunneling of electrons from the n-side to p-side takes place. The amount of the tunnel current is very large than the normal junction current. When the forward bias is increased, the tunnel current is increased up to certain limit.

When the band edge of n-side is same with the Fermi level in p-side the tunnel current is maximum with the further increment in the forward bias the tunnel current decreases and we get the desired negative conduction region. When the forward bias is raised further, normal pn junction current is obtained which is exponentially proportional to the applied voltage. The V-I characteristics of the tunnel diode is given.

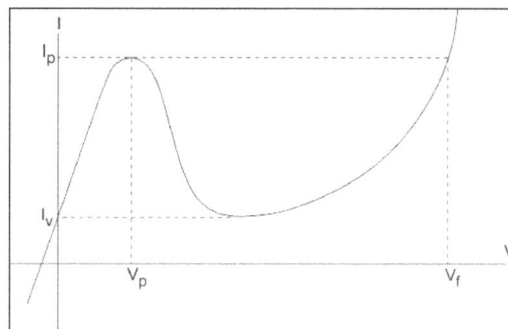

The negative resistance is used to achieve oscillation and often Ck+ function is of very high frequency frequencies.

Tunnel Diode Symbol

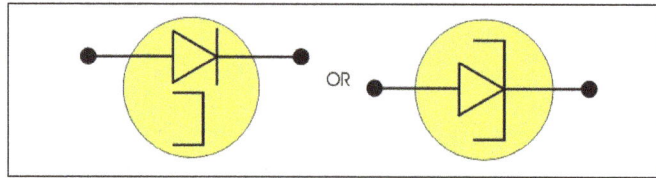

Tunnel Diode Applications

Tunnel diode is a type of sc diode which is capable of very fast and in microwave frequency range. It was the quantum mechanical effect which is known as tunneling. It is ideal for fast oscillators and receivers for its negative slope characteristics. But it cannot be used in large integrated circuits – that's why it's an applications are limited.

When the voltage is first applied current stars flowing through it. The current increases with the increase of voltage. Once the voltage rises high enough suddenly the current again starts increasing and tunnel diode stars behaving like a normal diode. Because of this unusual behavior, it can be used in number of special applications started below:

1. Oscillator Circuits: Tunnel diodes can be used as high frequency oscillators as the transition between the high electrical conductivity is very rapid. They can be used to create oscillation as high as 5Gz. Even they are capable of creativity oscillation up to 100 GHz in a appropriate digital circuits.

2. Used in Microwave Circuits: Normal diode transistors do not perform well in microwave operation. So, for microwave generators and amplifiers tunnel diode are. In microwave waves and satellite communication equipments they were used widely, but now a day's their uses is decreasing rapidly as transistor for working in wave frequency area available in market.

3. Resistant to Nuclear Radiation: Tunnel diodes are resistant to the effects of magnetic fields, high temperature and radioactivity. That's why these can be used in modern military equipment. These are used in nuclear magnetic resource machine also. But the most important field of its use satellite communication equipments.

Tunnel Diode Oscillator

Tunnel diode can make a very stable oscillator circuit when they are coupled to a tuned circuit or cavity, biased at the centre point of negative resistance region. Here is an example of tunnel diode oscillatory circuit.

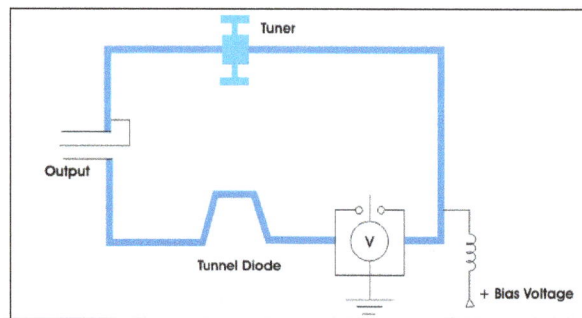

The tunnel diode is losing coupled to a tunable cavity. By using a short, antenna feed probe placed in the cavity off center loose coupling is achieved. To increase the stability of oscillation and achieve o/p power over wider bandwidth loose coupling is used. The range of the output power produced is few hundred micro-watts. This is useful for many microwave application. The physical position of the tuner determining the frequency of operation. If the frequency of operation is changed by this method, that is called mechanical tuning. Tunnel diode oscillators can be tuned electronically also.

Tunnel diode oscillators which are meant to be operated at microwave frequencies, generally used some form of transmission lines as tunnel circuit. These oscillators are useful in application that requires a few mill watts of power, example- local oscillators for microwave super electrodyne receiver.

Varactor Diode

Varactor Diode is a reverse biased p-n junction diode, whose capacitance can be varied electrically. As a result these diodes are also referred to as varicaps, tuning diodes, voltage variable capacitor diodes, parametric diodes and variable capacitor diodes. It is well known that the operation of the p-n junction depends on the bias applied which can be either forward or reverse in characteristic. It is also observed that the span of the depletion region in the p-n junction decreases as the voltage increases in case of forward bias. On the other hand, the width of the depletion region is seen to increase with an increase in the applied voltage for the reverse bias scenario.

Under such condition, the p-n junction can be considered to be analogous to a capacitor where the p and n layers represent the two plates of the capacitor while the depletion region acts as a dielectric separating them.

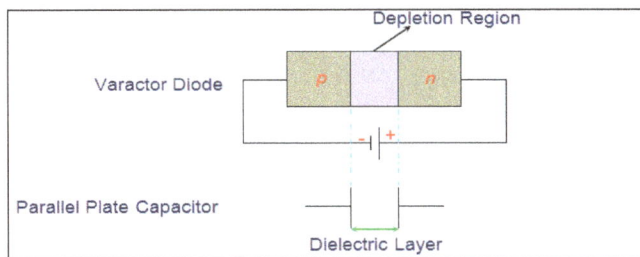

Varactor Diode Analogous to Parallel Plate Capacitor

Thus one can apply the formula used to compute the capacitance of a parallel plate capacitor even to the varactor diode.

Hence, mathematical expression for the capacitance of varactor diode is given by:

$$C_j = \frac{\varepsilon A}{d}$$

Where,

- C_j is the total capacitance of the junction.

- ε is the permittivity of the semiconductor material.

- A is the cross-sectional area of the junction.

- d is the width of the depletion region.

Further the relationship between the capacitance and the reverse bias voltage is given as:

$$C_j = \frac{CK}{\left(V_b - V_R\right)^m}$$

Where,

- C_j is the capacitance of the varactor diode.

- C is the capacitance of the varactor diode when unbiased.

- K is the constant, often considered to be 1.

- V_b is the barrier potential.

- V_R is the applied reverse voltage.

- m is the material dependent constant.

In addition, the electrical circuit equivalent of a varactor diode and its symbol are shown by figure. This indicates that the maximum operating frequency of the circuit is dependent on the series resistance (R_s) and the diode capacitance, which can be mathematically given as:

$$F = \frac{1}{2\pi R_s C_j}$$

In addition, the quality factor of the varactor diode is given by the equation:

$$Q = \frac{F}{f}$$

Where, F and f represent the cut-off frequency and the operating frequency, respectively.

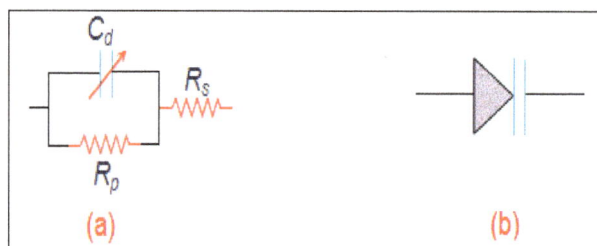

(a) Electrical Equivalent Circuit of Varactor Diode (b) Symbol of Varactor Diode

As a result, one can conclude that the capacitance of the varactor diode can be varied by varying the magnitude of the reverse bias voltage as it varies the width of the depletion region, d. Also it is evident from the capacitance equation that d is inversely proportional to C. This means that the junction capacitance of the varactor diode decreases with an increase in the depletion region width caused to due to an increase in the reverse bias voltage (V_R), as shown by the graph in figure. Meanwhile it is important to note that although all the diodes exhibit the similar property, varactor diodes are specially manufactured to achieve the objective. In other words varactor diodes are manufactured with an intention to obtain a definite C-V curve which can be accomplished by controlling the level of doping during the process of manufacture. Depending on this, varactor diodes can be classified into two types viz., abrupt varactor diodes and hyper-abrupt varactor diodes, depending on whether the p-n junction diode is linearly or non-linearly doped (respectively).

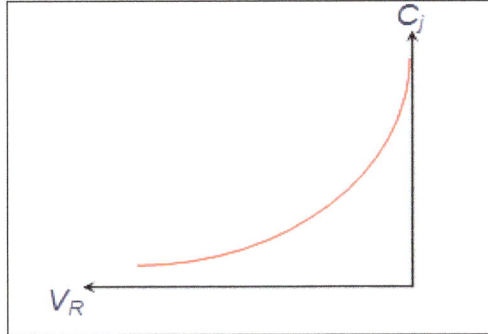

Characteristic Curve of a Varactor Diode

These varactor diodes are advantageous as they are compact in size, economical, reliable and less prone to noise when compared to other diodes. Hence, they are used in:

1. Tuning circuits to replace the old style variable capacitor tuning of FM radio,

2. Small remote control circuits,

3. Tank circuits of receiver or transmitter for auto-tuning as in case of TV,

4. Signal modulation and demodulation,

5. Microwave frequency multipliers as a component of LC resonant circuit,

6. Very low noise microwave parametric amplifiers,

7. AFC circuits,

8. Adjusting bridge circuits,

9. Adjustable bandpass filters,

10. Voltage Controlled Oscillators (VCOs),

11. RF phase shifters,

12. Frequency multipliers.

Schottky Diode

The name of this diode is given after the German physicist Walter. H. Schottky. Other than the name Schottky diode, it is also referred to as Schottky barrier diode or as hot carrier diode. This is a diode with semiconductor-metal junction.

This device can simply rectify frequencies greater than 300 MHz. Its forward voltage drop is also very low (0.15 to 0.45 V). This results in higher switching speed and improved system efficiency. The junction in the diode is formed by the metal (such as gold, tungsten, chromium, platinum, molybdenum or certain silicides) and N-type doped silicon semiconductor. Here, anode is the metal side and cathode is the semiconductor side.

The symbol of Schottky diode is in figure.

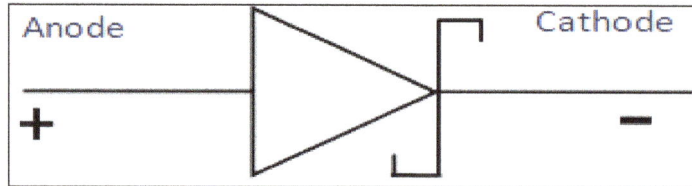

Construction of Schottky Diode

At one ending, there is a junction formed between the metal and lightly doped n-type semiconductor. This is unilateral junction. At the other ending, the metal and heavily doped semiconductor contact is present. It is called Ohmic bilateral contact. In this contact, no potential is present and is non-rectifying.

If the doping of semiconductor is increased, the depletion layer width decreases. When the width is decreased to a certain level, the charge carriers will tunnel easily through the depletion region. When the doping is very high, the junction can never act as a rectifier and it will become an ohmic contact. This diode can be a diode and an ohmic contact simultaneously.

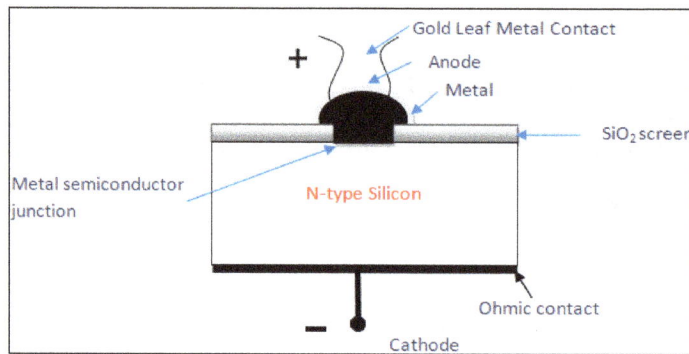

Passivated schottky Diode

When a Schottky diode is in unbiased condition, the electrons lying on the semiconductor side have very low energy level when compared to the electrons present in metal. Thus, the electrons cannot flow through the junction barrier which is called Schottky barrier. If the diode is forward biased, electrons present in the N-side gets sufficient energy to cross the junction barrier and enters into the metal. These electrons enter into the metal with tremendous energy. Consequently these electrons are known as hot carrier. Thus the diode is so called as hot-carrier diode.

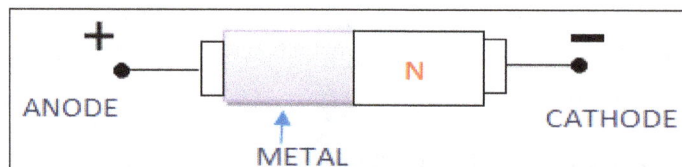

The equivalent circuit of the device (Schottky diode) with typical values of the components is shown in the figure.

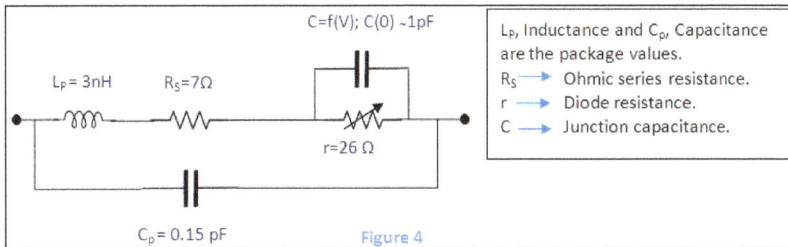

Figure 4

The above circuit can be approximated as shown below. This approximated circuit is used in many applications.

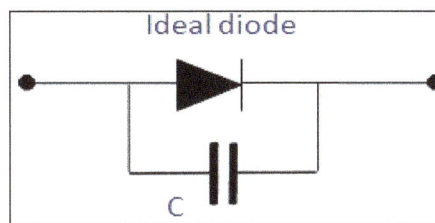

The Schottky diode has some unique features when compared to normal P-N junction diode:

- It is a unipolar device: This is due to the absence of significant current flow from metal to N-type semiconductor (minority carriers in the reverse direction is absent). But P-N junction diode is a bipolar device.

- No stored charge due to the absence of holes in the metal: As a result, schottky diode can quickly switch than other diodes and noise is also relatively low.

- Lower barrier potential (0.2 – 0.25 V) compared to P-N diode (0.7 V).

Comparison of V-I characteristics of Schottky Diode, PN Junction Diode and Point Contact Diode.

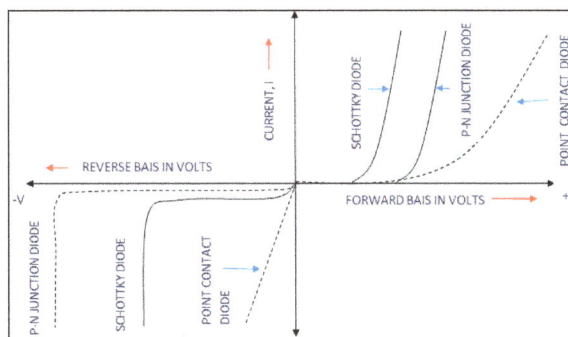

Comparison of V-I Characteristics of three Types of Diodes

Current Components in Schottky Doide

The current condition in this diode is through electrons (majority carriers) in N-type semiconductor.

$$I_T = I_{Diffusion} + I_{Tunneling} + I_{Thermonic\ emission}$$

Where,

- $I_{Tunneling} \rightarrow$ Tunneling current;

- $I_{Thermonic\ emission} \rightarrow$ As a result of electron ejection due to thermal energy (thermionic emission), this current will be produced across the electrodes;

- $I_{Diffusion} \rightarrow$ Diffusion current (result of concentration gradient $\left(\dfrac{dn}{dx}\right)$ and diffusion current density $\left(D_n \times q \times \dfrac{dn}{dx}\right)$ of electrons); where,

  $D_n \rightarrow$ Diffusion constant of electrons;

  $q \rightarrow$ Electronic charge = 1.6×10^{19} C.

Advantages of Schottky Diode

Advantages of Schottky diode are showing below:

- It has fast recovery time due to very low quantity of stored charge. So this diode is used for high speed switching application.

- It has low turn on voltage.

- It has low junction capacitance.

- Voltage drop is low.

Disadvantages of Schottky Diode

Disadvantages of Schottky diode are showing below:

- Reverse leakage current.

- Low reverse voltage rating.

Application of Schottky Diode

- Used in Switched-mode power supplies.

- Used in reverse current protection.

- Used in discharge protection.

- Used in voltage clamping application.

- Used in RF mixer and Detector diode.

- Used in solar cell application.

Photodiode

The photodiode is a kind of pn junction semiconductor diode which works with the intensity of light falling on it at the reverse biased condition.

Working Principle of Photodiode

When a diode is in reverse biased condition, there would be a reverse saturation current flowing through it from positive to the negative terminal of the diode. The unavoidable minority charge carriers cause this reverse saturation current in the semiconductor crystal. The value of this reverse saturation current does not depend on the applied reverse voltage across the diode rather it depends on the concentration of minority charge carriers in the semiconductor crystal. Hence for a certain range of reverse voltage across the diode, this current remains almost constant. We can control the reverse saturation current in a diode by controlling the concentration of minority charge carriers in the semiconductor crystal. We can change the concentration of minority charge carriers in a semiconductor by supplying external energy to the crystal.

In the photodiode, we do the same to control the conductivity of the device. As the name suggests in the photodiode, the pn junction gets exposed in the light. Depending on the intensity of the light, the covalent bonds in the crystal get broken and generate free electron-hole pairs across and nearby the pn junction. As a result, the reverse current in the diode gets increased or in other words the conductivity of the device increases.

Here it is to be noted that in a photodiode, only the pn junction portion of the diode must be exposed in light this is because if the light falls away from the junction, the electron-hole pairs created away from the junction get sufficient time to recombine hence they cannot contribute reverse current. But electron-hole pairs created in the junction or very nearby to the junction, can propagate easily towards opposite polarity due to the influence of electric field across the junction and hence the current through the photodiode gets increased.

Construction of Photodiode

The photodiodes are available in a metallic package. The diode is a p n junction, mounted in an insulated plastic substrate. Then we seal the plastic substrate in the metal case. On the top of the metal case, there is a transparent window, which allows light to entire up to the PN Junction. Two leads, anode and cathode of the diode come out from the bottom of the metal case. A tab extending from the side of the bottom portion of the metal case identifies the cathode lead.

Symbol of Photodiode

The symbol of a photodiode is just like ordinary diode except for two downward inclined arrows to symbolise the light.

Characteristic of Photodiode

1. Dark Resistance of Photodiode: It is true that there are always some minority charge carriers in the semiconductor crystal even in extreme dark condition — these minority charge carriers in the semiconductor crystal present due to unavoidable impurities and natural thermal excitation of the crystal. So even in dark condition, there would be a tiny and constant reverse saturation current in the diode. This current is fixed for a photodiode, and the current is known as dark current. The ratio of maximum withstandable reverse voltage to the dark current of a photodiode is called dark resistance of that diode.

$$R_d = \frac{\text{Maximum Reverse Voltage}}{\text{Dark Current}}$$

2. When we apply light to the diode, the reverse current increase. This relation is linear. The value of reverse current is directly proportional to the intensity of incident light energy.

3. If we go on increasing the light intensity, after a certain value of reverse current. The current will not increase further with increasing light intensity. We call this maximum value of reverse current as saturation current of the photodiode.

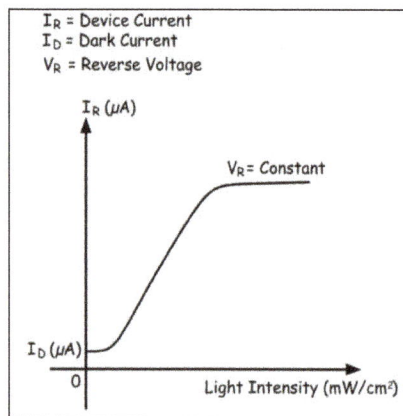

Applications of Photodiode

There are many applications of the photodiode. We don't have the scope of discussing all the applications of the photodiode. We shall discuss here only two popular application of photodiode.

Alarm Circuit Using Photodiode

Hence, we set a light source in such a way, the light always falls on a photodiode. As long as the light has strikes on the photodiode function, there would be a reverse current through the diode, as the diode is already connected in reverse biased condition in the alarm circuit. If an obstruction occurs in the light source, the reverse current in the photodiode comes down to dark current level. The circuit is so designed when the reverse current comes down, an alarm bell starts sounding. We may fit this arrangement in the doorway to defect the entry of people. In this arrangement, the light beam crosses the doorway from one side to another. We fit the photodiode in opposite side of the light source. When any person enters through the door, the light beam breaks and alarm sounds.

Counter Circuit using Photodiode

When numbers of items go through a conveyer belt, these can be easily counted by using photodiode. Here, we attach a light source on one side of the conveyer belt and a photodiode in opposite side of the belt. The light source and photodiode are fitted in such a way that, the light comes directly to the photodiode. As the light falls on the photodiode there will be reverse recovery current in the circuit. We connect the photodiode with a counter circuit. The light beam breaks, the counter gets one increment. When one item passes the light beam, it breaks the light beam and counter counts the item.

PIN Diode

PIN photodiode is a kind of photo detector; it can convert optical signals into electrical signals.

This technology was invented in the latest of 1950's. There are three regions in this type of diode. There is a p-region an intrinsic region and an n-region. The p-region and n-region are comparatively heavily doped than the p-region and n-region of usual p-n diodes. The width of the intrinsic region should be larger than the space charge width of a normal pn junction. The PIN photo diode operates with an applied reverse bias voltage and when the reverse bias is applied, the space charge region must cover the intrinsic region completely. Electron hole pairs are generated in the space charge region by photon absorption. The switching speed of frequency response of photodiode is inversely proportional to the life time.

The switching speed can be enhanced by a small minority carrier lifetime. For the photo detector applications where the speed of response is important, the depletion region width should be made as large as possible for small minority carrier lifetime as a result the switch speed also increases. This can be achieved PIN photodiode as the insertion of intrinsic region the space charge width larger. The diagram of a normal PIN photodiode is given below.

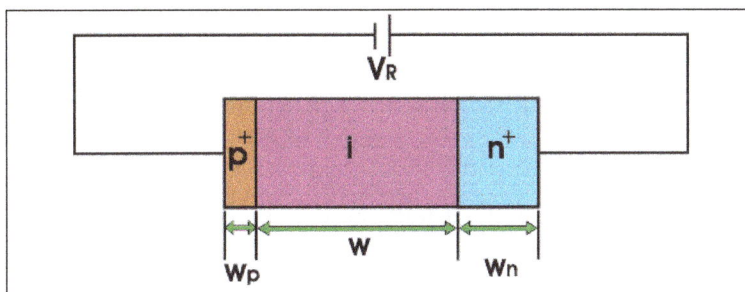

Avalanche Photo Diode

Avalanche photo diode is a kind of photo detector which can convert signals into electrical signals pioneering research work in the development of avalanche diode was done mainly in 1960's.

Avalanche photodiode structural configuration is very similar to the PIN photodiode. A PIN photodiode consists of three regions:

1. p-region,

2. intrinsic region,

3. n-region.

The difference is that reverse bias applied is very large to cause impact ionization. For silicon as the sc material, a diode will need between 100 to 200 volts. Firstly electron- hole pairs are generated by photon absorption in the depletion region. These generate more electron hole pairs through impact ionization. These are swept out of the depletion region quickly, i.e, the transit time is very less.

Laser Diode

Laser diodes are the semiconductor lasers which generate highly intense coherent beam of light. These were developed by Robert N. Hall in early 1960s and are also referred to as injection lasers. It is well known that an incident photon can interact with the atom to release a photon which will be identical to the impinging photon in all respects viz., phase, frequency, polarization and direction of travel. This phenomenon is referred to as stimulated emission and forms the basis of working for Lasers (Light Amplification by Stimulated Emission of Radiations). Further, if this event occurs in case of a p-n junction, then the diode is referred to as Laser diode. Laser Diodes are usually made of three layers (sometimes even two) where Gallium Arsenide (GaAs) like materials are doped with aluminium or silicon or selenium to produce p and n layers while the central, undoped, active layer is intrinsic in nature.

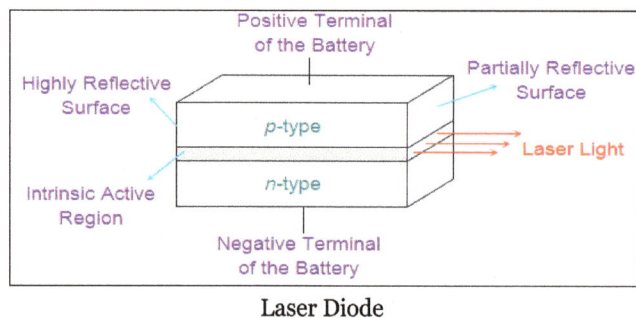

Laser Diode

When a large forward bias is applied for such an arrangement, heavy current flows through the junction due to which electrons will gain more energy when compared to holes. This extra energy is released in the form of photons when electrons combine with the holes (during recombination process). All these photons oscillate with a particular frequency and bounce back and forth between the reflective walls of the active layer. During this process, a few of them collide with the other atoms to produce more number of photons. This process continues and thus there will be an increase in the number of excited electrons when compared with those in the non-excited state.

This phenomenon is termed as population inversion and at this instant a constant highly coherent beam of light will be emitted from the central layer, in the direction parallel to it, through the partially reflecting surface as shown in figure.

Further it is to be noted that inorder to obtain laser light, the end surfaces of the semiconductor material should be parallel to each other, and are to be cut and polished carefully.

Among these, one is to be made fully-reflective in nature while the other should be partially-reflective. Also, the wavelength of the light emitted by the laser diode depends on the distance between these two reflecting surfaces and will usually lie in either visible or IR spectrum. This wavelength decides the size of the spot the laser light can be confined to i.e. shorter the wavelength smaller will be spot size. Laser diodes are compact in size and use little power and are thus preferred over lasers when the question will be of space and power consumption. Moreover laser diodes can be characterized in terms of their threshold current, Ith which indicates the value of current at which stimulated emission overcomes the spontaneous emission as shown by the L-I curve in figure.

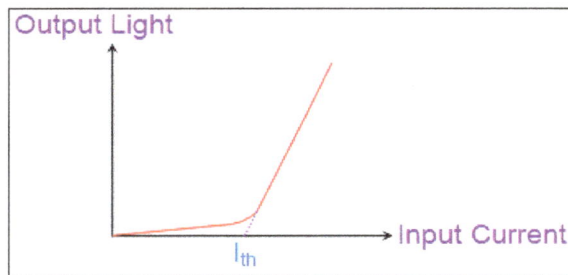

L-I Curve of Laser Diode

Laser diodes are available in wide variety of forms. If the active layer is narrow and if it can support only single optical mode of operation, resulting in a highly-focused beam, then such laser diodes are referred to as Single Mode Laser Diodes. On the other hand, Multi-Mode Laser Diodes have broader active region which supports multiple lateral optical modes and thus have high output power. However it is possible to obtain the output power greater than that of a single-mode laser diode without compromising on its confinement to narrow spectral range by using Master Oscillator Power Amplifier (MOPA) Laser Diodes, as they use an integrated power amplifier to increase the output power without affecting the narrow spectral output produced by the oscillator section. Further the laser diodes can either be Edge-emitting/Surface emitting depending on whether the laser light is emitted in the direction parallel or perpendicular to the material. Apart from these, there also exist many other variations of laser diodes like Vertical Cavity Surface Emitting Laser Diodes (VCSEL Diodes), High Power Laser Diodes, Stacked Laser Diodes, Distributed Feedback (DFB) Lasers or Distributed Bragg Reflector Lasers, External Cavity Diode Lasers, Broad Area Laser Diodes, Slab-Coupled Optical Waveguide Lasers (SCOWLs), etc.

Applications of Laser Diode:

1. CD and DVD players,

2. Barcode scanners,

3. Cable and High Definition (HD) TV transmission,

4. Medical applications including surgical instruments and to heal retina and brain,

5. Intrusion detection systems,

6. Remote control applications,

7. Industrial applications including welding, precision cutting of metals, heat treating, cladding, etc.,

8. Fibre Optic Communication,

9. High speed, Long distance communication,

10. Spectroscopic sensing,

11. Range finders,

12. Laser pointers,

13. Printing,

14. Integrated Circuits.

Light Emitting Diode

A Light Emitting Diode (LED) is a special type of PN junction diode. The light emitting diode is specially doped and made of a special type of semiconductor. This diode can emit light when it is in the forward biased state.

Aluminum indium gallium phosphide (AlInGaP) and indium gallium nitride (InGaN) are two of the most commonly used semiconductors for LED technologies. Older LED technologies used gallium arsenide phosphide (GaAsP), gallium phosphide (GaP), and aluminum gallium arsenide (AlGaAs).

LEDs generate visible radiation by electroluminescence phenomenon when a low-voltage direct current is applied to a suitably doped crystal containing a p-n junction, as shown in the diagram below. The doping is typically carried out with elements from column III and V of the periodic table. When a forward biased current, I_F, energizes the p-n junction, it emits light at a wavelength defined by the active region energy gap, e.g.

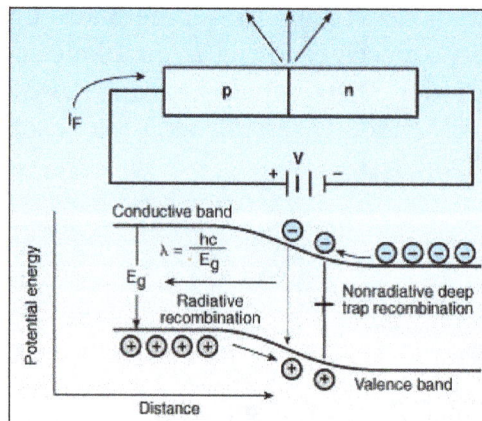

When the forward biased current I_F is applied through the p-n junction of the diode, minority carrier electrons are injected into the p-region and corresponding minority carrier electrons are injected into the n-region. Photon emission occurs due to electron-hole recombination in the p-region. Electron energy transitions across the energy gap, called radiative recombinations, produce photons (i.e., light), while shunt energy transitions, called non-radiative recombinations, produce phonons (i.e., heat). The luminous efficacies of typical AlInGaP LEDs and InGaN LEDs for different peak wavelengths are shown in the table.

	AlInGaP	InGaN
Energy gap (E_g)	1.8-2.31 eV	3.4 eV (blue)
Peak wavelength (λ)	585 nm (amber)	460 nm (blue)
		520nm (green)
Luminous efficacy (external)	20-25 lm/W (amber)	6 lm/W (blue)
		30 lm/W (green)

The efficacy depends on the light energy generated at the junction and losses due to re-absorption when light tries to escape through the crystal. The high index of refraction of most semiconductors causes the light to reflect back from the surface into the crystal and highly attenuated before finally exiting. The efficacy expressed in terms of this ultimate measurable visible energy is called the external efficacy.

The phenomenon of electroluminescence was observed in the year 1923 in naturally occurring junctions, but it was impractical at that time due to its low luminous efficacy in converting electric energy to light. But, today efficacy has increased considerably and LEDs are used not only in signals, indicators, signs, and displays but also in indoor lighting applications and road lighting applications.

Colour

The color of an LED device is expressed in terms of the dominant wavelength emitted, λd (in nm). AlInGaP LEDs produce the colors red (626 to 630 nm), red-orange (615 to 621 nm), orange (605 nm), and amber (590 to 592 nm). InGaN LEDs produce the colors green (525 nm), blue green (498 to 505 nm), and blue (470 nm). The color and forward voltage of AlInGaP LEDs depend on the temperature of the LED p-n junction. As the temperature of the LED p-n junction increases, the luminous intensity decreases, the dominant wavelength shifts towards longer wavelengths and the forward voltage drops. The variation in luminous intensity of InGaN LEDs with operating ambient temperature is small (about 10%) from − 20 °C to 80 °C. However, the dominant wavelength of InGaN LEDs does vary with LED drive current; as the LED drive current increases, dominant wavelength moves toward shorter wavelengths.

Dimming

LEDs may be dimmed to give 10% of their rated light output by reducing the drive current. LEDs are generally dimmed using Pulse Width Modulation techniques.

Reliability

The rated maximum junction temperature (TJMAX) is the most critical parameter for an LED. Temperatures exceeding this value usually result in damage of the plastic encapsulated LED device. Mean Time between Failures (MTBF) is used to find out the average life for LED. MTBF is determined by operating a quantity of LED devices at rated current in an ambient temperature of 55 °C and recording when half the devices fail.

White LEDs

White LEDs are being manufactured now using two methods: In the first method red, green, and blue LED chips are combined in the same package to produce white light; in the second method phosphorescence is used. Fluorescence in the phosphor that is encapsulated in the epoxy surrounding the LED chip is activated by the short-wavelength energy from the InGaN LED device.

Luminous Efficacy

Luminous efficacy of LED is defined as the emitted luminous flux (in lm) per unit electrical power consumed (in W). Blue LEDs have a rated internal efficacy in the order of 75 lm/W; red LEDs, approximately 155 lm/W; and amber LEDs, 500 lm/W. Taking into consideration losses due to internal re-absorption, the luminous efficacy is on the order of 20 to 25 lm/W for amber and green LEDs. This definition of efficacy is called external efficacy and is analogous to the definition of efficacy typically used for other light source types.

Transistor

The transistor is a semiconductor device which transfers a weak signal from low resistance circuit to high resistance circuit. The words trans mean transfer property and istor mean resistance property offered to the junctions. In other words, it is a switching device which regulates and amplify the electrical signal likes voltage or current.

The transistor consists two PN diode connected back to back. It has three terminals namely emitter, base and collector. The base is the middle section which is made up of thin layers. The right part of the diode is called emitter diode and the left part is called collector-base diode. These names are given as per the common terminal of the transistor. The emitter based junction of the transistor is connected to forward biased and the collector-base junction is connected in reverse bias which offers a high resistance.

Transistor Symbols

There are two types of transistor, namely NPN transistor and PNP transistor. The transistor which has two blocks of n-type semiconductor material and one block of P-type semiconductor material is known as NPN transistor. Similarly, if the material has one layer of N-type material and two layers of P-type material then it is called PNP transistor. The symbol of NPN and PNP is shown in the figure.

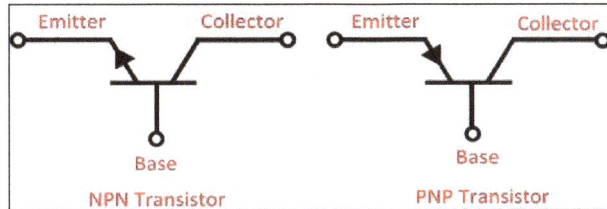

The arrow in the symbol indicates the direction of flow of conventional current in the emitter with forward biasing applied to the emitter-base junction. The only difference between the NPN and PNP transistor is in the direction of the current.

Transistor Terminals

The transistor has three terminals namely, emitter, collector and base. The terminals of the diode are explained below in details.

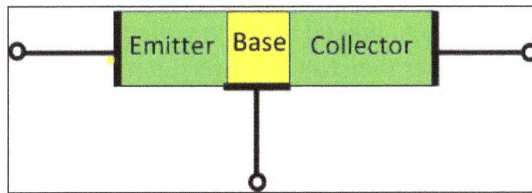

- Emitter – The section that supplies the large section of majority charge carrier is called emitter. The emitter is alway connected in forward biased with respect to the base so that it supplies the majority charge carrier to the base. The emitter-base junction injects a large amount of majority charge carrier into the base because it is heavily doped and moderate in size.

- Collector – The section which collects the major portion of the majority charge carrier supplied by the emitter is called a collector. The collector-base junction is always in reverse bias. Its main function is to remove the majority charges from its junction with the base. The collector section of the transistor is moderately doped, but larger in size so that it can collect most of the charge carrier supplied by the emitter.

- Base – The middle section of the transistor is known as the base. The base forms two circuits, the input circuit with the emitter and the output circuit with the collector. The emitter-base circuit is in forward biased and offered the low resistance to the circuit. The collector-base junction is in reverse bias and offers the higher resistance to the circuit. The base of the transistor is lightly doped and very thin due to which it offers the majority charge carrier to the base.

Working of Transistor

Usually, silicon is used for making the transistor because of their high voltage rating, greater current and less temperature sensitivity. The emitter-base section kept in forward biased constitutes the base current which flows through the base region. The magnitude of the base current is very small. The base current causes the electrons to move into the collector region or create a hole in the base region.

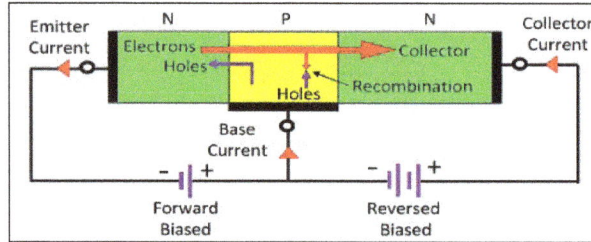

Working of Transistor

The base of the transistor is very thin and lightly doped because of which it has less number of electrons as compared to the emitter. The few electrons of the emitter are combined with the hole of the base region and the remaining electrons are moved towards the collector region and constitute the collector current. Thus we can say that the large collector current is obtained by varying the base region

Transistor Operating Conditions

When the emitter junction is in forward biased and the collector junction is in reverse bias, then it is said to be in the active region. The transistor has two junctions which can be biased in different ways. The different working conduction of the transistor is shown in the table.

Condition	Emitter Junction (EB)	Collector Junction (CB)	Region of Operation
FR	Forward-biased	Reversed-biased	Active
FF	Forward-biased	Forward-biased	Saturation
RR	Reversed-biased	Reversed-biased	Cut-off
RF	Reversed-biased	Forward-biased	Inverted

- FR – In this case, the emitter-base junction is connected in forward biased and the collector-base junction is connected in reverse biased. The transistor is in the active region and the collector current is depend on the emitter current. The transistor, which operates in this region is used for amplification.

- FF – In this condition, both the junction is in forward biased. The transistor is in saturation and the collector current becomes independent of the base current. The transistors act like a closed switch.

- RR – Both the current are in reverse biased. The emitter does not supply the majority charge carrier to the base and carriers current are not collected by the collector. Thus the transistors act like a closed switch.

- RF – The emitter-base junction is in reverse bias and the collector-base junction is kept in forward biased. As the collector is lightly doped as compared to the emitter junction it does not supply the majority charge carrier to the base. Hence poor transistor action is achieved.

Bipolar Junction Transistor

The Bipolar Junction Transistor is a semiconductor device which can be used for switching or amplification. Bipolar transistors have the ability to operate within three different regions:

- Active Region – The transistor operates as an amplifier and $Ic = \beta*Ib$.

- Saturation – The transistor is "Fully-ON" operating as a switch and $Ic = I(saturation)$.

- Cut-off – The transistor is "Fully-OFF" operating as a switch and $Ic = 0$.

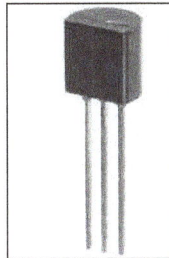

A Typical Bipolar Transistor

The Bipolar Transistor basic construction consists of two PN-junctions producing three connecting terminals with each terminal being given a name to identify it from the other two. These three terminals are known and labelled as the Emitter (E), the Base (B) and the Collector (C) respectively.

Bipolar Transistors are current regulating devices that control the amount of current flowing through them from the Emitter to the Collector terminals in proportion to the amount of biasing voltage applied to their base terminal, thus acting like a current-controlled switch. As a small current flowing into the base terminal controls a much larger collector current forming the basis of transistor action.

The principle of operation of the two transistor types PNP and NPN, is exactly the same the only difference being in their biasing and the polarity of the power supply for each type.

Bipolar Transistor Construction

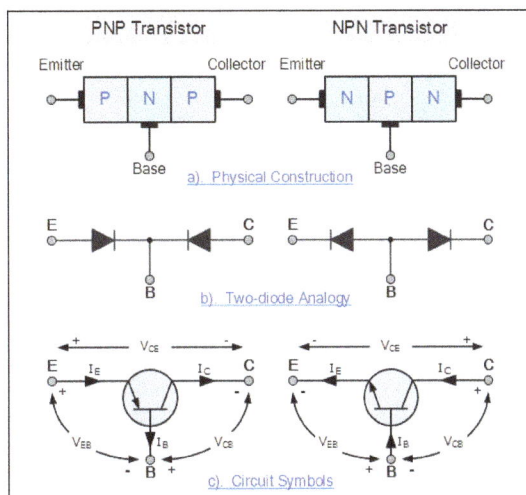

The construction and circuit symbols for both the PNP and NPN bipolar transistor are given above with the arrow in the circuit symbol always showing the direction of "conventional current flow" between the base terminal and its emitter terminal. The direction of the arrow always points from the positive P-type region to the negative N-type region for both transistor types, exactly the same as for the standard diode symbol.

Types of BJT

NPN Transistor

NPN transistor is one of the Bipolar Junction Transistor (BJT) types. The NPN transistor consists of two n-type semiconductor materials and they are separated by a thin layer of p-type semiconductor. Here the majority charge carriers are the electrons. The flowing of these electrons from emitter to collector forms the current flow in the transistor. Generally the NPN transistor is the most used type of bipolar transistors because the mobility of electrons is higher than the mobility of holes. The NPN transistor has three terminals – emitter, base and collector. The NPN transistor is mostly used for amplifying and switching the signals.

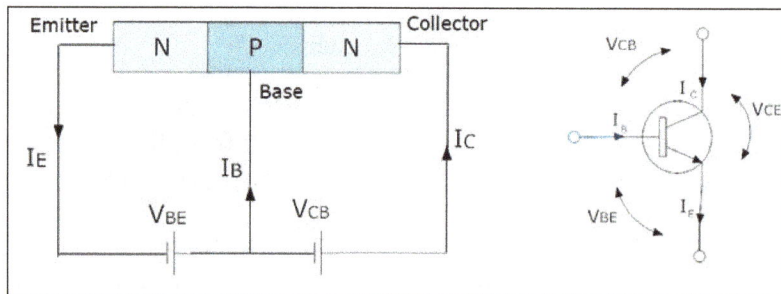

The figure shows the symbol and structure of NPN transistor. In this structure we can observe the three terminals of transistor, circuit currents and voltage value representations.

NPN Transistor Circuit

The figure shows the NPN transistor circuit with supply voltages and resistive loads. Here the collector terminal always connected to the positive voltage, the emitter terminal connected to the negative supply and the base terminal controls the ON/OFF states of transistor depending on the voltage applied to it.

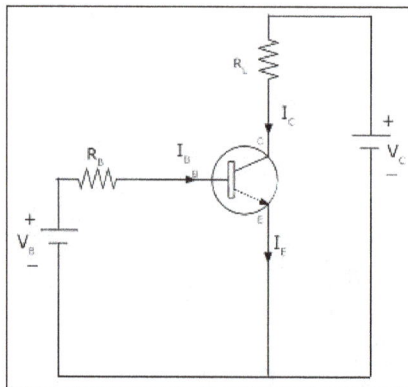

NPN Transistor Working

The working of NPN transistor is quite complex. In the above circuit connections we observed that the supply voltage VB is applied to the base terminal through the load RB. The collector terminal connected to the voltage VCC through the load RL. Here both the loads RB and RL can limit the current flow through the corresponding terminals. Here the base terminal and collector terminals always contain positive voltages with respect to emitter terminal.

If the base voltage is equal to the emitter voltage then the transistor is in OFF state. If the base voltage increases over emitter voltage then the transistor becomes more switched until it is in fully ON state. If the sufficient positive voltage is applied to the base terminal i.e. fully-ON state, then electrons flow generated and the current (IC) flows from emitter to the collector. Here the base terminal acts as input and the collector-emitter region acts as output.

To allow current flow between emitter and collector properly, it is necessary that the collector voltage must be positive and also greater than the emitter voltage of transistor. Some amount of voltage drop presented between base and emitter, such as 0.7V. So the base voltage must be greater than the voltage drop 0.7V otherwise the transistor will not operate. The equation for base current of a bipolar NPN transistor is given by:

$$I_B = (V_B - V_{BE}) / R_B$$

Where,

I_B = Base current,

V_B = Base bias voltage,

V_{BE} = Input Base-emitter voltage = 0.7V,

R_B = Base resistance.

The output collector current in common emitter NPN transistor can be calculated by applying Kirchhoff's Voltage Law (KVL).

The equation for collector supply voltage is given as:

$$V_{CC} = I_C R_L + V_{CE}$$

From the above equation the collector current for common emitter NPN transistor is given as:

$$I_C = (V_{CC} - V_{CE}) / R_L$$

In a common emitter NPN transistor the relation between collector current and emitter current is given as:

$$I_C = \beta I_B$$

In active region the NPN transistor acts as a good amplifier. In common emitter NPN transistor total current flow through the transistor is defined as the ratio of collector current to the base current IC/IB. This ratio is also called as "DC current gain" and it doesn't have any units. This ratio is generally represented with β and the maximum value of β is about 200. In common base NPN transistor the total current gain is expressed with the ratio of collector current to emitter current IC/IE. This ratio is represented with α and this value is generally equal to unity.

α, β and γ Relationship in NPN Transistor

Now let us see the relationship between the two ratio parameters α and β.

α = DC current gain for common base circuit = Output current/Input current

In common base NPN transistor output current is collector current (IC) and input current is emitter current (IE).

$$\alpha = I_C / I_E$$

This current gain (α) value is very close to unity but less than the unity.

The emitter current is the sum of small base current and large collector current.

$$I_E = I_C + I_B$$
$$I_B = I_E - I_C$$

From $\alpha = I_C / I_E$, the collector:

$$I_C = \alpha I_E$$
$$I_B = I_E - \alpha I_E$$
$$I_B = I_E (1-\alpha)$$

β = DC current gain for common emitter circuit = Output current/Input current

Here output current is collector current and input current is base current.

$$\beta = I_C / I_B$$
$$\beta = I_C / I_E (1-\alpha)$$
$$\beta = \alpha / (1-\alpha)$$

From the above equations the relationship between α and β can be expressed as:

$$\alpha = \beta (1-\alpha) = \beta / (\beta+1)$$
$$\beta = \alpha (1+\beta) = \alpha / (1-\alpha)$$

The β value may vary from 20 to 1000 for low power transistors which operate with high frequencies.

But in general this β value can have the values in between the range of 50-200.

The relationship between α, β and γ factors. In common collector NPN transistor the current gain is defined as the ratio emitter current IE to base current IB. This current gain is represented with γ.

$$\gamma = I_E / I_B$$

Emitter current:

$$I_E = I_C + I_B$$
$$\gamma = (I_C + I_B)/I_B$$
$$\gamma = (I_C / I_B) + 1$$
$$\gamma = \beta + 1$$

Hence the relationships between α, β and γ are given as:

$$\alpha = \beta / (\beta + 1), \ \beta = \alpha / (1 - \alpha), \ \gamma = \beta + 1$$

Common Emitter Configuration

The common emitter configuration circuit is one of the three BJT configurations. These common emitter configuration circuits are used as voltage amplifiers. Generally the BJT transistors have three terminals but in the circuit connections we need to take any one terminal as common. So we use one of the three terminals as common terminal for both input and output actions. In this configuration we use emitter terminal as common terminal, thus it is named as common-emitter configuration.

This configuration is used as a single stage common emitter amplifier circuit. In this configuration base acts as input terminal, collector acts as output terminal and emitter as common terminal. The operation of this circuit starts with biasing the base terminal such that forward biasing the base-emitter junction. The small current in the base controls the current flow in the transistor. This configuration always operates in the linear region to amplify the signals at output side.

This common emitter amplifier gives the inverted output and can have very high gain. This gain value is influenced by temperature and bias current. The common-emitter amplifier circuit is mostly used configuration than other BJT configurations because of its high input impedance and low output impedance and also this configuration amplifier provides high voltage gain and power gain.

The current gain for this configuration is always greater than unity usually the typical value is about 50. These configuration amplifiers are mostly used in the applications where low frequency amplifier and radio frequency circuits are required. The circuit diagram for the common-emitter amplifier configuration is shown in the figure.

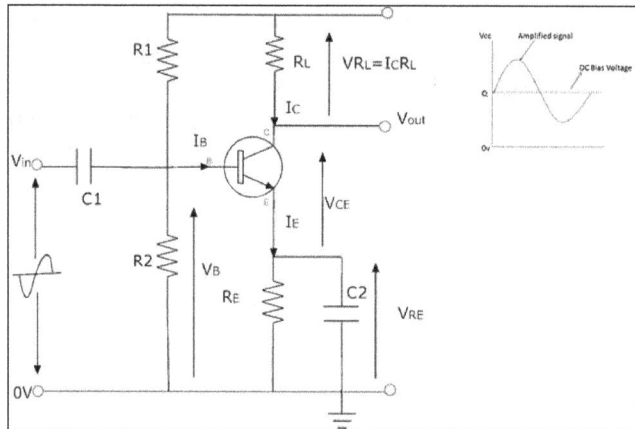

Output Characteristics of NPN Transistor

The family of output characteristics curves of a bipolar transistor is given below. The curves show the relationship between the collector current (IC) and the collector-emitter voltage (VCE) with the varying of base current (IB). The transistor is 'ON' only when at least a small amount of current and small amount of voltage is applied at its base terminal relative to emitter otherwise the transistor is in 'OFF' state.

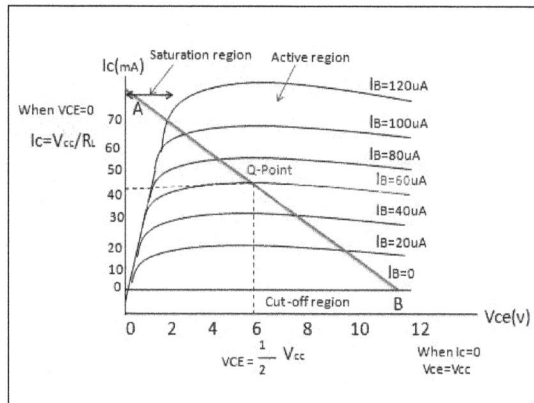

The collector current (IC) is mostly affected by the collector voltage (VCE) at 1.0V level but this IC value is not highly affected above this value. Already the emitter current is the sum of base and collector currents. i.e. IE = IC + IB. The current flowing through the resistive load (RL) is equal to the collector current of the transistor. The equation for the collector current is given by:

$$I_C = (V_{CC} - V_{CE}) / R_L$$

The straight line indicates the 'Dynamic load line' which is connecting the points A (where $V_{CE}= 0$) and B (where $I_C = 0$). The region along this load line represents the 'active region' of the transistor.

The common emitter configuration characteristics curves are used to calculate the collector current when the collector voltage and base current is given. The load line (red line) is used to determine the Q-point in the graph. The slope of the load line is equal to the reciprocal of the load resistance. i.e. -1/RL.

NPN Transistor Applications

- NPN transistors are mainly used in switching applications.

- Used in amplifying circuit applications.

- Used in the Darlington pair circuits to amplify weak signals.

- NPN transistors are used in the applications where there is a need to sink a current.

- Used in some classic amplifier circuits, such as 'push-pull' amplifier circuits.

- In temperature sensors.

- Very High frequency applications.

- Used in logarithmic convertors.

PNP Transistor

PNP transistor is another type of Bipolar Junction Transistor (BJT). The structure of the PNP transistor is completely different from the NPN transistor. The two PN-junction diodes in the PNP transistor structure are reversed with respect to the NPN transistor, such as the two P-type doped semiconductor materials are separated by a thin layer of N-type doped semiconductor material.

In PNP transistor, the majority current carriers are holes and electrons are the minority current carriers. All the supply voltage polarities applied to the PNP transistor are reversed. In PNP, the current sinks in to the base terminal. The small base current in the PNP has the ability to control the large emitter-collector current because it is a current-controlled device.

The arrow for BJT transistors is always located on the emitter terminal and also it indicates the direction of conventional current flow. In PNP transistor this arrow indicates as 'pointing in' and the current direction in PNP is completely opposite to the NPN transistor. The structure of PNP transistor is completely opposite to the NPN transistor. But the characteristics and operation of the PNP transistor is almost same as NPN transistor with small differences. The symbol and structure for PNP transistor is shown in the figure.

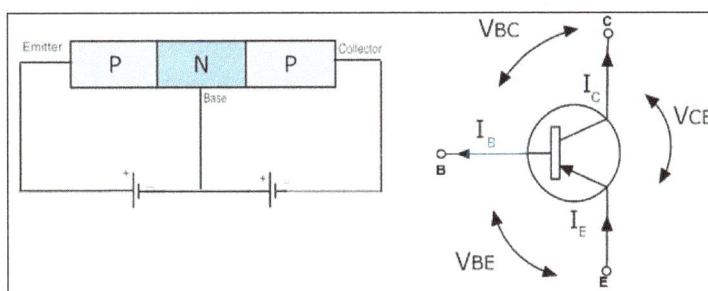

The figure shows the structure and symbol of PNP Transistor. This transistor mainly consists of 3 terminals and they are Emitter (E), Collector (C) and Base (B). Here if you observe, the base current flows out of the base unlike NPN transistor. The emitter voltage is much positive with respect to base and collector.

PNP Transistor Working

The circuit connection of PNP transistor with supply voltages is given below. Here the base terminal has negative bias with respect to emitter and the emitter terminal has positive bias voltage with respect to both base and collector because of PNP transistor.

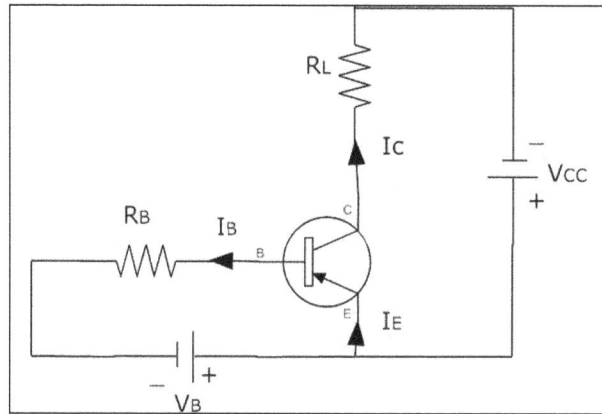

The polarities and current directions are reversed here compared to NPN transistor. If the transistor is connected to all the voltage sources as shown above then the base current flows through the transistor but here the base voltage needs to be more negative with respect to the emitter to operate transistor. Here the base- emitter junction acts as a diode. The small amount of current in the base controls the flowing of large current through emitter to collector region. The base voltage is generally 0.7V for Si and 0.3V for Germanium devices.

Here the base terminal acts as input and the emitter- collector region acts as output. The supply voltage V_{cc} is connected to the emitter terminal and a load resistor (R_L) is connected to the collector terminal. This load resistor (R_L) is used to limits the maximum current flow through the device. One more resistor (R_B) is connected to the base terminal which is used to limit the maximum current flow through the base terminal and also a negative voltage is applied to the base terminal. Here the collector current is always equal to the subtraction of base current from emitter current. Like NPN transistor, the PNP transistor also has the current gain value β. Now let us see the relation between the currents and current gain β.

The collector current (I_c) is given by:

$$I_C = I_E - I_B$$

The DC current gain (β) for the PNP transistor is same as the NPN transistor.

DC current gain = β = Output current/Input current

Here output current is collector current and input current is base current.

$$\beta = I_C / I_B$$

From this equation we get:

$$I_B = I_C / \beta$$
$$I_C = \beta I_B$$

And also we define the current gain as:

Current gain = Collector current/Emitter current (In common base transistor)

$$\alpha = I_C / I_E$$

The relation between α and β is given by:

$$\beta = \alpha / (1 - \alpha) \text{ and } \alpha = \beta / (\beta + 1)$$

The collector current in PNP transistor is given by:

$$I_C = - \alpha I_E + I_{CBO}$$

Where,

I_{CBO} is the saturation current.

Since $I_E = -(I_C + I_B)$

$$I_C = -\alpha \left(-(I_C + I_B) \right) + I_{CBO}$$
$$I_C - \alpha I_C = \alpha I_B + I_{CBO}$$
$$I_C (1 - \alpha) = \alpha I_B + I_{CBO}$$
$$I_C = \left(\alpha / (1 - \alpha) \right) I_B + I_{CBO} / (1 - \alpha)$$

Since $\beta = \alpha / (1 - \alpha)$

Now we get the equation for collector current:

$$I_C = \beta I_B + (1 + \beta) I_{CBO}$$

The output characteristics of PNP transistor are same as NPN transistor characteristics. The small difference is that the PNP transistor characteristic curve rotates 180° to calculate the reverse polarity voltages and current values. The dynamic load line also exists on the characteristic curve to calculate the Q-point value. The PNP transistors are also used in switching and amplifying circuits like NPN transistors.

PNP Transistor Example

Consider a PNP transistor, which is connected in the circuit with the supply voltages $V_B = 1.5V$, V_E

= 2V, $+V_{CC} = 10V$ and $-V_{CC} = -10V$. And also this circuit connected with the resistors of $R_B = 200k\Omega$ and $R_E = R_C$ (or R_L) = 5kΩ. Now calculate the current gain values (α, β) of the PNP transistor.

Here,

$$V_B = 1.5V$$
$$V_E = 2V$$
$$+V_{CC} = 10V \text{ and } -V_{CC} = -10V$$
$$R_B = 200k\Omega$$
$$R_E = R_C(\text{or } R_L) = 5k\Omega$$

Base current:

$$I_B = V_B / R_B = 1.5 / (200^*10^3) = 7.5uA.$$

Emitter current:

$$I_E = V_E / R_E = (10-2)/(5^*10^3) = 8/(5^*10^3) = 1.6mA.$$

Collector current:

$$I_C = I_E - I_B = 1.6^*10^{-3} - 7.5^*10^{-6} = 1.59mA.$$

Now we have to calculate α and β values:

$$\alpha = I_C / I_E = 1.59^*10^{-3} / 1.6^*10^{-3} = 0.995$$
$$\beta = I_C / I_B = 1.59^*10^{-3} / 7.5^*10^{-6} = 212$$

Finally we get the current gain values of a considered PNP transistor are:

$$\alpha = 0.995 \text{ and } \beta = 212.$$

BJT Transistor Matching

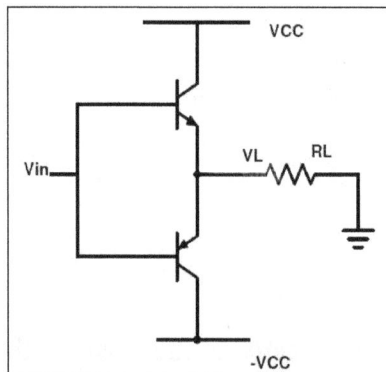

Transistor matching is nothing but connecting the both NPN and PNP transistors in a single design to generate high power. This structure is also called as "matched pair". The both NPN and PNP transistors are called complementary transistors. Mainly these matched pair circuits are used in power amplifiers, such as class B amplifiers. If we connect the complementary transistors which are having the same characteristics then it is very useful to operate the output stages in motors and large machinery designs by producing high power continuously.

The NPN transistor conducts only in the positive half cycle of signal and the PNP transistor conducts only in the negative half cycle of the signal, due to this the device operates continuously. This continuous operation is very useful in the power motors to produce continuous power. The complementary transistors need to have same DC current gain (β) value. These matched pair circuits are used in motors controlling, robotics and power amplifier applications.

PNP Transistor Identification

Generally we identify the PNP transistors with their structure. We have some differences in the structures of both NPN and PNP transistors when compared. One more thing to identify the PNP transistor is generally the PNP transistor is in OFF for positive voltage and it is in ON when small output current and negative voltage at its base with respect to emitter. But to identify them with most efficiently we use some other technique by calculating the resistance between the three terminals, such as base, emitter and collector.

We have some standard resistance values for identifying the both NPN and PNP transistors. It is necessary to test each pair of terminals in both directions for resistance values so totally six tests are needed. This process is very much useful to identify the PNP transistor easily. Now we see the operation behavior of each pair of terminals:

- Emitter-Base Terminals: The emitter-base region acts as a diode but it conducts only in one direction.

- Collector-Base Terminals: The collector-base region also acts as diode which conducts I only one direction.

- Emitter-Collector Terminals: The emitter- collector region looks like a diode but it will not conduct in either direction.

Now let us see the resistance value table to identify both NPN and PNP transistors as shown in the following table.

Between Transistor Terminals		PNP	NPN
Collector	Emitter	R_{HIGH}	R_{HIGH}
Collector	Base	R_{LOW}	R_{HIGH}
Emitter	Collector	R_{HIGH}	R_{HIGH}
Emitter	Base	R_{LOW}	R_{HIGH}
Base	Collector	R_{HIGH}	R_{LOW}
Base	Emitter	R_{HIGH}	R_{LOW}

PNP Transistor as a Switch

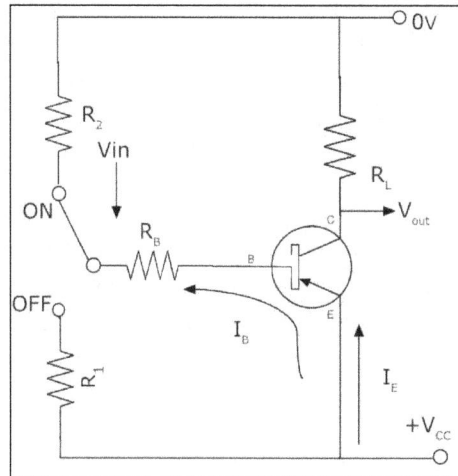

The circuit in the figure shows the PNP transistor as a switch. The operation of this circuit is very simple, if the input pin of transistor (base) is connected to ground (i.e. negative voltage) then the PNP transistor is in 'ON', now the supply voltage at emitter conducts and the output pin pulled up to the larger voltage. If the input pin connected to the high voltage (i.e. positive voltage) then the transistor is 'OFF', so the output voltage has to be low (zero). This operation shows the switching conditions of a PNP transistor due to their ON and OFF states.

Applications

- PNP transistors are used to source current, i.e. current flows out of the collector.

- PNP transistors are used as switches.

- These are used in the amplifying circuits.

- PNP transistors are used when we need to turnoff something by push a button, i.e. emergency shutdown.

- Used in Darlington pair circuits.

- Used in matched pair circuits to produce continuous power.

- Used in heavy motors to control current flow.

- Used in robotic applications.

Field Effect Transistor

The name field effect transistor itself hints the attributes of this device. Like BJT it is also capable of transferring a signal from high resistance to low resistance (Transfer resistor-transistor). The term field effect stands for the fact that the operation of the device mainly depends on the electric field applied between its terminals called Gate and source analogous to the base and collector in transistor to control the current through it. That is the electrical field applied between two terminals (Gate and source Vgs) controls the flow of current in other terminal (Drain).

Types of Field Effect Transistors

Field Effect transistor's are of mainly of two types based on construction features of the device:

- Junction Field Effect Transistor.

- Metal oxide semiconductor Field Effect Transistor.

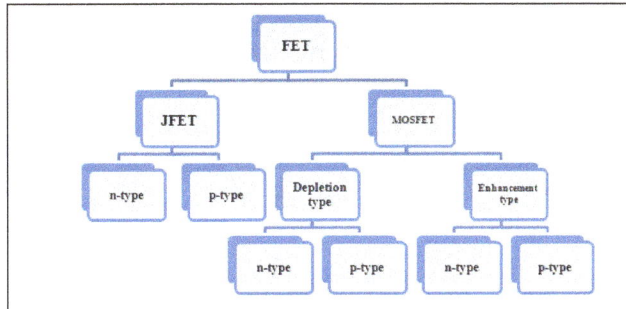

Types of field effect transistors

JFET's are again subdivided into two types based on the type of channel namely p channel JFET and n- channel JFET. Similarly MOSFET's are divided into two types a) Enhancement mode MOSFET 2) Depletion mode MOSFET. In Enhancement mode MOSFET channel is induced by applying gate voltage contrary to the depletion mode MOSFET in which already existing channel is modulated by applying gate voltage like JFET.

Types of JFET

There are two types of JFET based on the type of channel i.e. based on the majority carrries flow:

- N- channel JFET

- P- channel JFET

The circuit symbol of P channel JFET and N channel JFET is as shown in the figure with the arrow on gate terminal shows the device polarity. Since JFET is a symmetric device either ends of N-channel can be used as source or drain. It is useful for the circuit design to show either of two terminals as source and other as drain. It is achieved here by placing gate terminal close to source.

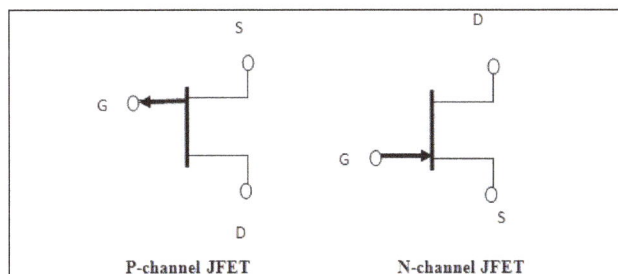

N channel and P channel FET symbols

JFET Structure

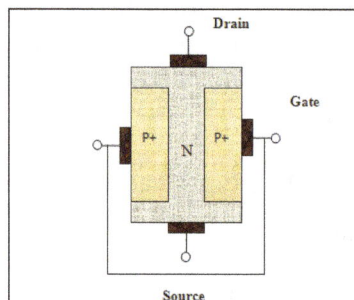

JET Structure

N-channel JFET consists of a p+ type semiconductors grown by doping acceptor impurities on either side of N-type semiconductor as shown in the figure.

Current is allowed to flow through the length of the N-channel between the two p+ semiconductors. FET is a three terminal devicewith the terminals being source, gate and drain.

Source

It is there terminal where the majority carriers enter in to JFET bar. As for the FET bar concern this terminal is the source of majority carriers so it is called as source terminal and the current flow in this terminal is I_s.

Drain

This terminal is the end terminal which collects the majority carriers sourced by the source. I.e. it is draining the majority carriers from FET bar and giving to the output terminal so it is called as drain terminal and the current in drain is I_d.

Gate

This important control element in FET as it acts like gate to the majority carriers flow. I.e. by operating the gate terminal voltage opening or closing of majority carries can be obtained, so this terminal called Gate terminal and the current in gate is I_g.

Practically FET is manufactured by doping donor impurities in p-type substrate. So along with these three terminals there exists substrate terminal which is often shorted to source. Both terminals of p+ type semiconductor are joined to form gate terminal. Metal contacts are provided for all the three terminals. Electrons flow from source to drain hence the conventional current of positive charge carriers flows from drain to source.

MOSFET

The MOSFET or metal oxide semiconductor field effect transistor is a form of FET that offers exceedingly high input impedance.

The gate input has an oxide layer insulating it from the channel and as a result its input resistance is very many MΩ.

The MOSFET has a number of different characteristics compared to the junction FET, and as a result it can be used in a number of different areas and it is able to provide excellent performance.

One particular area where MOSFET technology is used is within CMOS logic integrated circuits. The extraordinarily high input impedance means that these circuits are able to consume very low power levels and this means that high levels of integration can be achieved.

A typical discrete MOSFET in a plastic encapsulation

MOSFET Circuit Symbol

There is a variety of different circuit symbols used for MOSFETs. In view of the different variety of standards used, along with the different types of MOSFET, a host of different MOSFET circuit symbols can be seen.

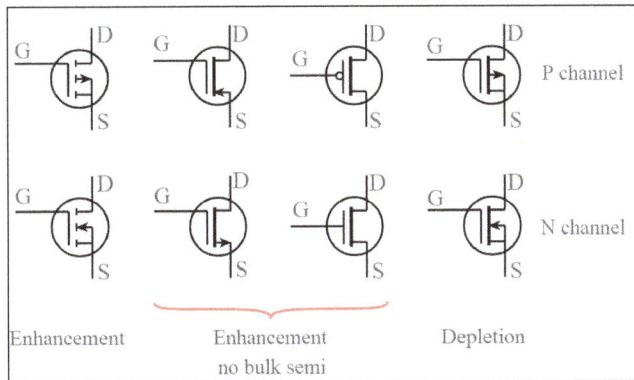

MOSFET circuit symbols

- The circuit symbol for the basic MOSFET (shown left most) indicates that the device has a bulk substrate - this is indicated by the arrow on the central area of the substrate.

- The MOSFET circuit symbols shown in the centre and marked "enhancement no bulk semi" are both valid and used equally frequently. They indicate an enhancement MOSFET that has no bulk semiconductor.

- Depletion mode MOSFETs are generally indicated as shown on the right most section.

MOSFET circuit symbols for both P-channel and N-channel types are shwn. The drain is shown at the top as this is generally as they are seen on circuit diagrams.

MOSFET Key Parameters

Before looking at the operation of the MOSFET, it is worth summarising a few of the key features associated with MOSFET technology.

Key MOSFET Features	
Feature	Details
Gate construction	The gate is physically insulated from the channel by an oxide layer. Voltages applied to the gate control the conductivity of the channel as a result of the electric field induced capacitively across the insulating dielectric layer.
N / P channel	Both N-channel and P-channel variants are available.
Enhancement / depletion	Both enhancement and depletion types are available. As the name suggests the depletion mode MOSFET acts by depleting or removing the current carriers from the channel, whereas the enhancement type increases the number of carriers according to the gate voltage.

MOSFETs may be characterised as N Channel and P-Channel. Each has different characteristics:

Comparison of the Key Features of N-Channel and P-Channel MOSFETS		
Parameter	N-Channel	P-Channel
Source / drain material	N-Type	P-Type
Channel material	P-Type	N-Type
Threshold voltage V_{th}	negative	doping dependent
Substrate material	P-Type	P-Type
Inversion layer carriers	Electrons	Holes

MOSFET Operation

Like other forms of FET, the current flowing in the channel of the MOSFET is controlled by the voltage present on the gate. As such MOSFETs are widely used in applications such as switches and also amplifiers. They are also able to consume very low levels of current and as a result they are widely used in microprocessors, logic integrated circuits and the like. CMOS integrated circuits used MOSFET technology.

Also like other forms of FET, the MOSFET is available in depletion mode and enhancement mode variants. The enhancement mode is what may be termed normally OFF, i..e when the VGS gate source voltage is zero and requires a gate voltage to turn it on, whereas the other form, deletion mode devices are normally ON when VGS is zero.

There are basically three regions in which MOSFETs can operate:

- Cut-off region: In this region of the MOSFET is in a non-conducting state, i.e. turned OFF - channel current I_{DS}= 0. The gate voltage V_{GS} is less than the threshold voltage required for conduction.

- Linear region: In this linear region the channel is conducting and controlled by the gate voltage. For the MOSFET to be in this state the V_{GS} must be greater than the threshold voltage and also the voltage across the channel, V_{DS} must be greater than V_{GS}.

- Saturation region: In this region the MOSFET is turned hard on. The voltage drop for a MOSFET is typically lower than that of a bipolar transistor and as a result power MOSFETs are widely used for switching large currents.

A power MOSFET in a TO220 package

Switching for Different Types of MOSFET			
MOSFET Type	V_{GS} +VE	V_{GS} 0	V_{GS} -VE
N-Channel Enhancement	ON	OFF	OFF
N-Channel Depletion	ON	ON	OFF
P-Channel Enhancement	OFF	OFF	ON
P-Channel Depletion	OFF	ON	ON

MOSFET Structure

As already implied the key factor of the MOSFET is the fact that the gate is insulated from the channel by a thin oxide layer. This forms one of the key elements of its structure.

For an N-channel device the current flow is carried by electrons and in the diagram below it can be seen that the drain and source are formed using N+ regions which provide good conductivity for these regions.

In some structures the N+ regions are formed using ion implantation after the gate area has been formed. In this way, they are self-aligned to the gate.

The gate to source and gate to drain overlap are required to ensure there is a continuous channel. Also the device is often symmetrical and therefore source and drain can be interchanged. On some higher power designs this may not always be the case.

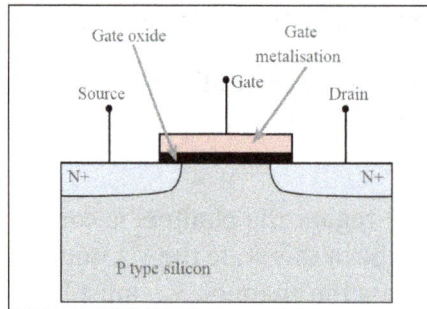

N channel enhancement mode MOSFET structure

It can be seen from the diagram that the substrate is the opposite type to the channel, i.e. P-type rather than N-type, etc. This is done to achieve source and drain isolation.

The oxide over the channel is normally grown thermally as this ensures good interfacing with the substrate and the most common gate material is polysilicon, although some metals and silicides can be used.

The depletion mode has a slightly different structure. For this a separate N-type channel is set up within the substrate.

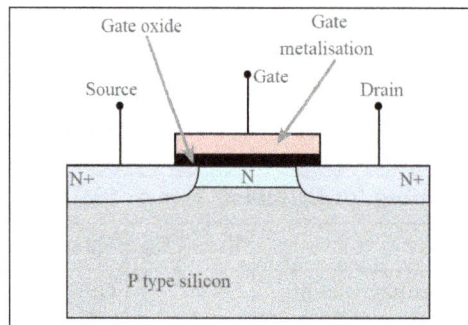

N channel depletion mode MOSFET structure

P-channel FETs are not as widely used. The main reason for this is that the holes do not have as high a level of mobility as electrons, and therefore the performance is not as high. However they are often required for use in complementary circuits, and it is mainly for this reason that they are manufactured or incorporated into ICs.

MOSFETs are possibly the most widely used active device. As they appear in CMOS and other integrated circuit technologies where they enable very low power operation - a requirement for very large scale integration, otherwise power consumption would be far too high.

Not only are they used in IC technology, but they are also used as discrete components as well where they are able to offer very high input impedance levels and also low noise operation in oscillators, amplifiers and many other circuits.

Dual Gate MOSFET

The dual gate MOSFET is a useful form of MOSFET which can provide some distinct advantages, especially in RF applications.

The dual gate MOSFET can be considered in the same light as the tetrode vacuum tube or thermionic valve. The introduction of the second control electrode considerably reduced the level of feedback capacitance between the input and output circuits of the device. In this way it is possible to make far more stable amplifiers.

Also the additional gate enables the dual gate MOSFET to be used in RF mixer or RF multiplier circuits.

Dual Gate MOSFET Circuit Symbol

The circuit symbol from the dual gate MOSFET expands the basic single gate MOSFET and adds a second gate into the input.

Enhancement and depletion mode as well as N channel and P channel devices can be described, although P channel devices tend not to be used much for RF applications because hole mobility is much less than electron mobility.

The basic dual gate MOSFET circuit symbols are shown in the figure.

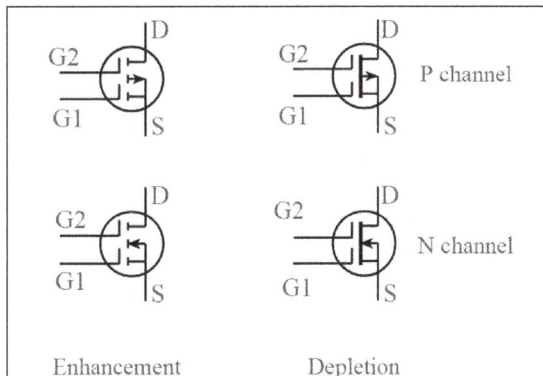

MOSFET circuit symbols

The dual gate MOSFET can be used in a number of applications including RF mixers /multipliers, RF amplifiers, amplifiers with gain control and the like.

Dual Gate MOSFET Structure

The dual gate MOSFET has what may be referred to as a tetrode construction where the two grids control the current through the channel.

The different gates control different sections of the channel which are in series with each other.

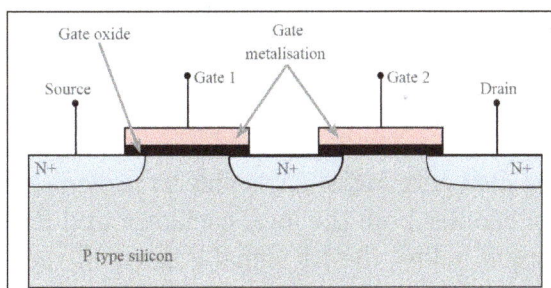

Dual gate MOSFET structure

Dual Gate MOSFET Applications

Dual gate MOSFETs are used in many applications. Their attributes and characteristics mean that they can provide some distinct advantages for some forms of circuit.

1. RF amplifier: Dual gate MOSFETs are able to operate with improved performance as amplifiers over single gated FETs. The dual gate MOSFET enables a cascode two stage amplifier to be constructed using a single device.

The cascade amplifier helps overcome the Miller effect where capacitance is present between the input and output stages. Although the Miller effect can relate to any impedance between the input and output, normally the most critical is capacitance. This capacitance can lead to an increase in the level of input capacitance experienced and in high frequency (e.g. VHF & UHF) amplifiers it can also lead to instability.

The effect is overcome by using a cascade amplifier using a single dual gate FET. In this configuration, biasing the drain-side gate at constant potential reduces the gain loss caused by Miller effect. The effects of capacitive coupling between the input and output are virtually eliminated.

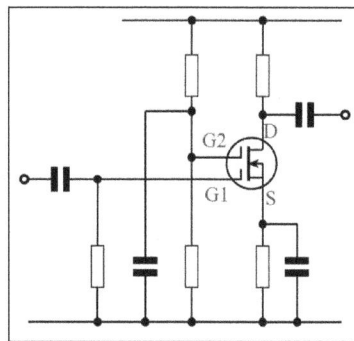

Basic dual gate MOSFET amplifier circuit

In this circuit the lower or input FET section is in a self-biased, common-source configuration. The upper or output FET section is configured in a in a voltage-divider biased, common-gate configuration.

Effectively a cascade amplifier is a two-stage amplifier formed from a trans-conductance amplifier which is followed by a current buffer. This provides a high level of input-output isolation, high input impedance, high output impedance, higher gain or higher bandwidth when compared to a single stage amplifier.

A cascode amplifier using a dual gate MOSFET is commonly used in radio receiver front ends. In these applications, the dual-gate MOSFET is operated as a common source amplifier with the primary gate, i.e. gate 1, G1 connected to the input and the second gate, G2 grounded to RF via the capacitor.

2. RF mixer / multiplier: The dual gate MOSFET is able to provide a basis for an RF mixer. The dual gate MOSFET operation enables both the local oscillator and RF signal inputs to be accommodated. As shown in the circuit below, the RF signal is normally applied to gate 1 and the local oscillator to gate 2.

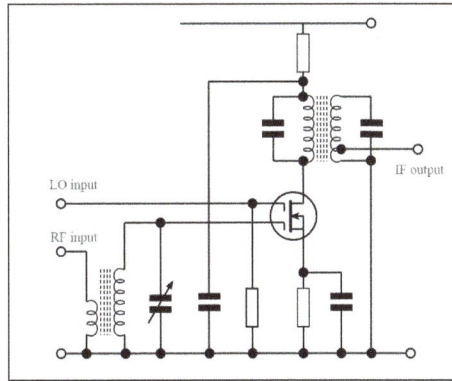

Basic dual gate MOSFET mixer circuit

The operation of this dual gate MOSFET circuit is relatively easy to understand. The RF signal appears at gate 1 and controls the channel current in the normal way. However the much higher level local oscillator signal is applied to gate 2 and superimposes its effect on the channel current.

3. Level / gain control: The output from the dual gate MOSFET is proportional to the input at both of the gates. With a constant level at gate 1, for example, varying the voltage on gate 2 will alter the output level. Accordingly the dual gate MOSFET can be used to provide linear gain control.

The dual gate MOSFET is a useful component to be able to incorporate into circuit designs when appropriate. Although single gate FETs are by far the most widely used, the characteristics of the dual gate MOSFET can provide some very useful improvements in performance in some applications.

Power MOSFET

Power MOSFTs are used in many power supply and general power applications, especially as switches. Variants include planar MOSFETs, VMOS, UMOS TrenchMOS, HEXFETs and other different brand names.

MOSFET technology is ideal for use in many power applications, where the low switch on resistance enables high levels of efficiency to be attained.

There is a number of different varieties of power MOSFET available from different manufacturers, each with its own characteristics and abilities.

Many power MOSFETs incorporate a vertical structure topology. This enables high current switching with high efficiency within a relatively small die area. It also enables the device to support high current and voltage switching.

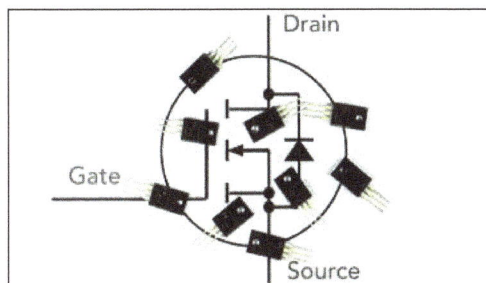

Power MOSFET Types

Within the overall arena of power MOSFETs, there are a number of specific technologies that have been developed and addressed by different manufacturers. They use a number of different techniques that enable the power MOSFETs to carry the current and handle the power levels more efficiently. As already mentioned they often incorporate a form of vertical structure.

The different types of power MOSFET have different attributes and therefore can be particularly suited for given applications:

- Planar power MOSFET: This is the basic form of power MOSFET. It is good for high voltage ratings because the ON resistance is dominated by the epi-layer resistance. This structure is generally used when a high cell density is not needed.

- VMOS: VMOS power MOSFETs have been available for many years. The basic concept uses a V groove structure to enable a more vertical flow of the current, thereby providing lower ON resistance levels and better switching characteristics. Although used for power switching, they may also be used for high frequency small RF power amplifiers.

- UMOS: The UMOS version of the power MOSFET uses a grove similar to that the VMOS FET. However the grove has a flatter bottom to it and provides some different advantages.

- HEXFET: This form of power MOSFET uses a hexagonal structure to provide the current capability.

- TrenchMOS: Again the TrenchMOS power MOSFET uses a similar basic grove or trench in the basic silicon to provide better handling capacity and characteristics. In particular, Trench power MOSFETs are mainly used for voltages above 200 volts because of their channel density and hence their lower ON resistance.

Power MOSFET Breakdown Voltage

The breakdown voltage is a key parameter for any power device including power MOSFETs. As these devices may operate a voltages well in excess of those encountered in lower power electronic circuits, the voltage breakdown voltage is an important aspect of any power MOSFET device.

In most power MOSFETs the N+ source termination and the P body junction are shorted using source metallisation. This avoids the possibility of spurious turn on of the parasitic bipolar transistor within the structure.

In operation, when no bias is applied to the gate, then the device is able to provide a high drain voltage through the reverse biased P type body and N+ epitaxial layer junction (shown as P-silicon and N- on the planar power MOSFET diagram). When high voltages are present, most of the applied voltage appears across the lightly doped N- layer. If a higher operational voltage is required, then the N- layer can be more lightly doped and made thicker, but this also has the effect of increasing the ON resistance.

For lower voltage devices, the doping levels for the P silicon areas and the N- become comparable and the voltage is shared across these two layers. However if the P silicon area is not thick enough then it can be found that the depletion region can punch through to the N+ source region, giving rise to a lower breakdown voltage.

On the other hand, if the device is designed for too high a voltage, then the channel resistance and threshold voltage will increase. As a result careful optimisation of the device is needed. Also when choosing power MOSFET devices, it is necessary to opt for one that provides the correct combination of breakdown voltage and ON resistance.

A power MOSFET in a TO220 package

Capacitance

The switching behaviour of any power MOSFET is greatly affected by the levels of parasitic capacitance that occurs within the device.

The main areas of capacitance that affect the switching performance are gate to source capacitance C_{GS}; gate to drain capacitance, C_{GD}; and the drain to source, C_{DS}.

These capacitances are non-linear and they are dependent upon the device structure and the voltages present at any given time. Thy result from the bias dependent oxide capacitance and the bias dependent depletion layer capacitance. Typically as the voltages increase, so the depletion layers increase and the capacitance levels decrease.

Power MOSFET Threshold Voltage

The threshold voltage which is normally designated as $V_{GS(TH)}$ is the minimum gate voltage that can form a conducting channel between the source and the drain.

For power MOSFETs this threshold voltage is normally measured for a drain source current of 250μA.

The threshold voltage is determined by factors in the power MOSFET including the gate oxide thickness and the doping concentration in the channel.

Power MOSFET Applications

Power MOSEFET technology is applicable to many types of circuit. Applications include:

- Linear power supplies

- Switching power supplies

- DC-DC converters

- Low voltage motor control.

Power MOSFETs are normally used in applications where voltages do not exceed about 200 volts. Higher voltages are not so easily achievable. Where the Power MOSFETs are used, it is their low ON resistance that is particularly attractive. This reduces power dissipation which reduces cost and size less metalwork and cooling is required. Also the low ON resistance means that efficiency levels can be maintained at a higher level.

MOSFET Applications

MOSFET as Switch

A direct consequence of MOSFET working leads to their usage as a switch. A n-channel MOSFET shown by figure can act as a switching circuit when it operates in cut-off and saturation regions. This is because the MOSFET in the figure will be ON when the V_{GS} voltage is positive, which causes the MOSFET to behave like a short circuit. Thus one gets the output voltage almost equal to zero. Further if the V_{GS} is zero, then the MOSFET will be OFF and thus the output voltage will be equal to V_{DD}.

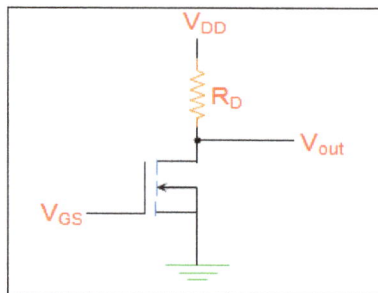

MOSFET as a switch

Such MOSFETs are used to perform switching actions in case of basic buck converters used in DC-DC power supplies. Here one MOSFET switch stores the energy into the inductor while, the other releases it into the load, in alternate cycles.

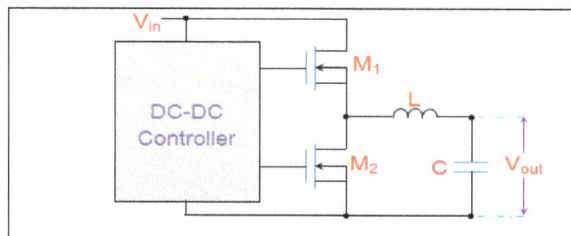

MOSFETs used to switch the power supply

Further the MOSFETs are used even in the case of motor control applications, where the switching action is controlled by either a half-bridge control circuit which uses two MOSFETs or by a full-bridge control circuit using four MOSFETs. Here the movement of DC motors or brushless stepper motors can be controlled by using Power MOSFETs by employing the techniques like Pulse Width Modulation (PWM).

MOSFET as Amplifiers

Enhancement n-channel MOSFETs are in their OFF state when no gate-to-source voltage is applied. However when biased with a suitable positive voltage, it starts to conduct allowing the flow of drain current through it. This current is seen to increase in magnitude as the bias voltage increases which in turn leads to the increase in output voltage. Thus the MOSFETs serve as amplifiers as shown by figure. MOSFET amplifiers are used in radio-frequency applications and in sound systems.

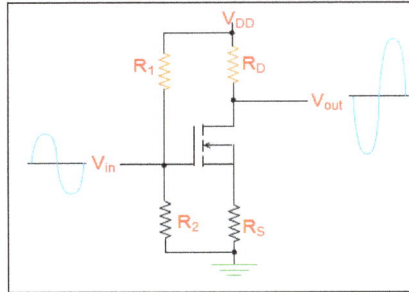

MOSFET as an amplifier

MOSFET as Chopper

The switching action of MOSFETs can be exploited to design chopper circuits as shown by figure. Here the DC voltage, V_{DC} is converted into AC voltage with the same amplitude level, V_{AC} by biasing the MOSFET using a square voltage waveform between its gate and source terminals. This causes the MOSFETs to operate in cut-off and saturation regions in alternate cycles.

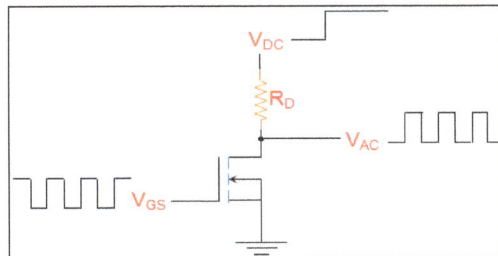

MOSFET as a chopper

MOSFET as Linear Voltage Regulators

Depletion type MOSFETs in source-follower configuration are used in linear voltage regulator circuits as pass transistors. Here the source voltage, V_L follows the gate voltage, V_G minus the gate-to-source voltage, V_{GS}. Further V_{GS} increases with an increase in the drain current, I_D. Thus if the gate voltage is fixed, then the source voltage will reduce as the load current, I_L increases.

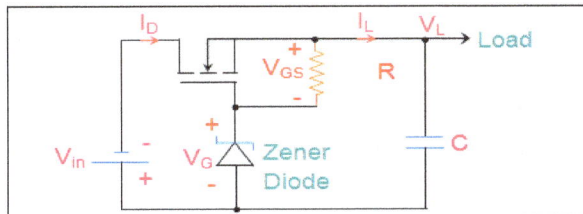

MOSFET as a linear voltage regulator

Apart from these, Depletion mode MOSFETs are used in the design of off-line switch-mode power supplies which operate under the scenario of wide voltage variations. Further they are used in the design of variable frequency drives, voltage-ramp generator circuits and serve as voltage-controlled potentiometers due to their property to behave as voltage-controlled resistors. MOSFETs can also be used as constant current sources and can be used to design current monitoring circuits in conjunction with op-amps. They are used as mixers or oscillators in radio systems and can be used to drive high-current/high voltage networks without drawing current or power from the driving circuit, due to their high impedance which isolates the two parts of the network. The MOSFETs also find their applications in the fields of digital electronics and microprocessors, due to their high input impedance and fast switching speeds.

Thyristor

The thyristor is a form of electronics component which consists of four layers of differently doped silicon rather than the three layers of the conventional bipolar transistors.

Whereas conventional transistors may have a p-n-p or n-p-n structure with the electrodes named collector, base and emitter, the thyristor has a p-n-p-n structure with the outer layers with their electrodes referred to as the anode (n-type) and the cathode (p-type). The control terminal of the SCR is named the gate and it is connected to the p-type layer that adjoins the cathode layer.

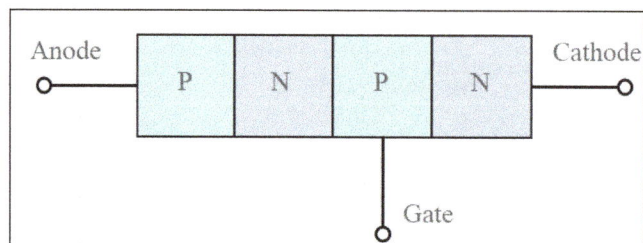

Basic structure of a thyristor/SCR

Thyristors are usually manufactured from silicon, although, in theory other types of semiconductor could be used. The first reason for using silicon for thyistors is that silicon is the ideal choice because of its overall properties. It is able to handle the voltage and currents required for high power applications. Additionally it has good thermal properties. The second major reason is that silicon technology is well established and it is widely used for a variety of semiconductor electronics components. As a result it is very cheap and easy for semiconductor manufacturers to use.

Thyristor Symbols and Basics

The thyristor or silicon controlled rectifier, SCR is a device that has a number of unusual characteristics. It has three terminals: Anode, cathode and gate, reflecting thermionic valve / vacuum tube technology. As might be expected the gate is the control terminal while the main current flows between the anode and cathode.

The device is a "one way device" giving rise to the GE name for it the silicon controlled rectifier. Therefore when the device is used with AC, it will only conduct for a maximum of half the cycle.

In operation, the thyristor or SCR will not conduct initially. It requires a certain level of current to flow in the gate to "fire" it. Once fired, the thyristor will remain in conduction until the voltage across the anode and cathode is removed - this obviously happens at the end of the half cycle over which the thyristor conducts. The next half cycle will be blocked as a result of the rectifier action. It will then require current in the gate circuit to fire the SCR again.

The silicon controlled rectifier, SCR or thyristor symbol used for circuit diagrams or circuit seeks to emphasis its rectifier characteristics while also showing the control gate. As a result the thyristor symbol shows the traditional diode symbol with a control gate entering near the junction.

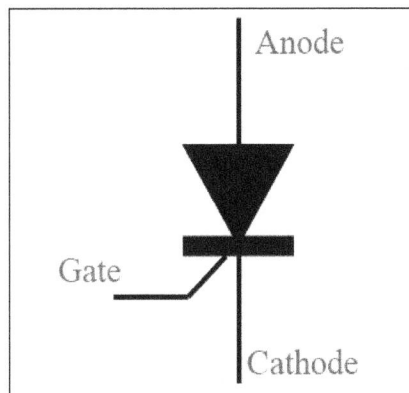

Thyristor or SCR circuit symbol

Thyristor Specifications

In order to select the correct thyristor device for any circuit, it is necessary to study the datasheets and ensure that the device has the right characteristics for the intended circuit or application.

Thyristors are rather unique components and their specifications and datasheet parameters are different to the more widely used transistors and FETs, etc.

Other Types of Thyristor or SCR

There is a number of different type's thyristor - these are variants of the basic component, but they offer different capabilities that can be used in various instances and may be useful for certain circuits.

- Reverse conducting thyristor, RCT: Although thyristors normally block current in the reverse direction, there is one form called a reverse conducting thyristor which has an integrated reverse diode to provide conduction in the reverse direction, although there is no control in this direction.

  Within a reverse conducting thyristor, the device itself and the diode do not conduct at the same time. This means that they do not produce heat simultaneously. As a result they can be integrated and cooled together.

The RCT can be used where a reverse or freewheel diode would otherwise be needed. Reverse conducting thyristors are often used in frequency changers and inverters.

- Gate Assisted Turn-Off Thyristor, GATT: The GATT is used in circumstances where a fast turn-off is needed. To assist in this process a negative gate voltage can sometimes be applied. In addition to reducing the anode cathode voltage. This reverse gate voltage helps in draining the minority carriers stored on the n-type base region and it ensures that the gate-cathode junction is not forward biased.

  The structure of the GATT is similar to that of the standard thyristor, except that the narrow cathode strips are often used to enable the gate to have more control because it is closer to the centre of the cathode.

- Gate Turn-Off Thyristor, GTO: The GTO is sometimes also referred to as the gate turn off switch. This device is unusual in the thyristor family because it can be turned off by simply applying a negative voltage to the gate - there is no requirement to remove the anode cathode voltage.

- Asymmetric Thyristor: This device is used in circuits where the thyristor does not see a reverse voltage and therefore the rectifier capability is not needed. As a result it is possible to make the second junction, often referred to as J2.

Two Transistor Analogy of Thyristor

Here, the equivalent circuit of two transistor shows that the base of PNP transistor T1 is fed by collector current of NPN transistor T2 and collector current of transistor T1 feeds base of transistor T2. Hence, conduction of both the transistor depends on each other. So, until one of the base of any transistor get base current it will not conduct even if the voltage is present at anode and cathode. The main difference between the transistor and Thyristor is, transistor turns off as the base current is removed while the Thyristor remains ON just by trigger it once. For applications like alarm circuit which need to trigger once and stay ON forever, cannot use transistor. So, for overcome these problems we use Thyristor.

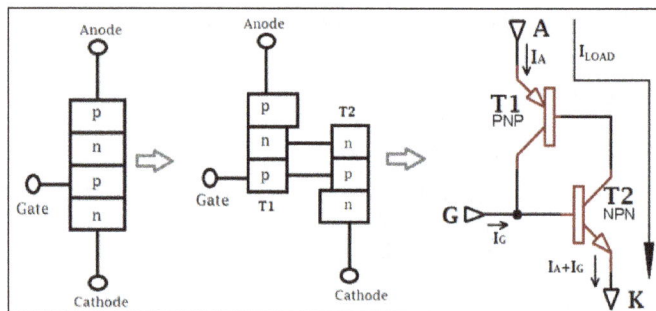

V-I Characteristics of Thyristor or SCR

The basic circuit for obtaining Thyristor V-I characteristics is given below, the anode and cathode of the Thyristor are connected to main supply through the load. The gate and cathode of the Thyristor are fed from a source Es, used to provide gate current from gate to cathode.

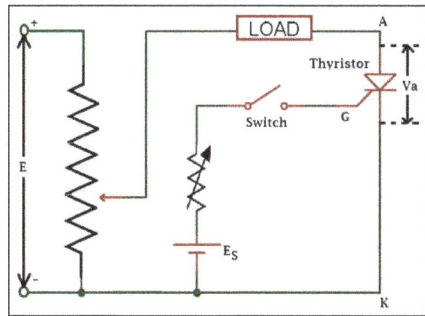

Basic circuit for obtaining V-I characteristic

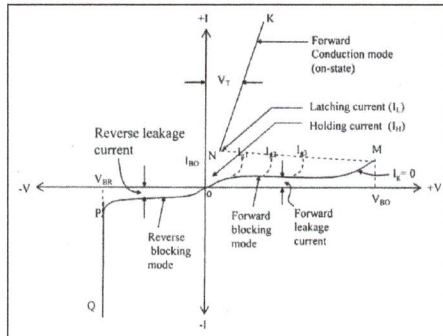

V-I characteristics

As per the characteristic diagram, there are three basic modes of SCR: reverse blocking mode, forward blocking mode, and forward conduction mode.

Reverse Blocking Mode

In this mode the cathode is made positive with respect to anode with switch S open. Junction J1 and J3 are reversed biased and J2 is forward biased. When reverse voltage applied across Thyristor (should be less than V_{BR}), the device offers a high impedance in the reverse direction. Therefore, Thyristor treated as open switch in the reverse blocking mode. V_{BR} is the reverse breakdown voltage where the avalanche occurs, if voltage exceeds V_{BR} may cause to Thyristor damage.

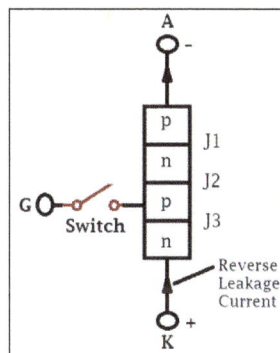

Forward Blocking Mode

When anode is made positive with respect to cathode, with gate switch open. Thyristor is said to be forward biased, junction J1 and J3 are forward biased and J2 is reversed biased as you can see in figure. In this mode, a small current flows called forward leakage current, as the forward leakage current is small and not enough to trigger the SCR. Therefore, SCR is treated as open switch even in forward blocking mode.

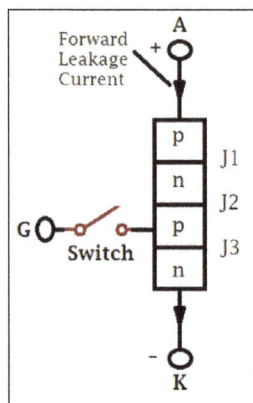

Forward Conduction Mode

As the forward voltage is increased with gate circuit remain open, an avalanche occurs at junction J2 and SCR comes into conduction mode. We can turn ON the SCR at any moment by giving a positive gate pulse between gate and cathode or by a forward breakover voltage across anode and cathode of the Thyristor.

Triggering Methods of SCR or Thyristor

There are many methods to triggering the SCR like:

- Forward Voltage Triggering
- Gate Triggering
- dv/dt triggering
- Temperature Triggering
- Light Triggering

Forward Voltage Triggering

By applying forward voltage between anode and cathode, with keeping gate circuit open, junction J2 is reverse biased. As a result, the formation of depletion layer occurs across J2. As the forward voltage increases, a stage comes when depletion layer get vanishes, and J2 is said to have Avalanche Breakdown. Hence, Thyristor comes in conduction state. The voltage at which the avalanche occurs called as forward breakover voltage V_{BO}.

Gate Triggering

It is one of the most common, reliable and efficient way to turning ON the Thyristor or SCR. In gate triggering, to turning ON an SCR, a positive voltage is applied between gate and cathode, which gives rise to the gate current and the charge gets injected into the inner P layer and forward brea-kover occurs. As higher the gate current will lower the forward breakover voltage.

As shown in figure there are three junction in a SCR, now for turning ON the SCR the junction J2 should breaks. By using gate triggering method, as the gate pulse applied the junction J2 breaks, junction J1 and J2 gets forward biased or the SCR comes in conduction state. Hence, it allows the current to flow through anode to cathode.

As per the two transistor model, when the anode is made positive with respect to cathode. Current will not flow through anode to cathode until the gate pin is triggered. When current flows into the gate pin it turns ON the lower transistor. As lower transistor conduct, it turns ON the upper tran-sistor. This is a kind internal positive feedback, so by providing pulse at gate for one time, made the Thyristor to stay in ON condition. When both the transistor turns ON current start conducting through anode to cathode. This state is known as forward conducting and this is how a transistor "latches" or stays permanently ON. For turning OFF the SCR, you cannot off it just by removing gate current, at this state Thyristor get independent of gate current. So, for turning OFF you have to make switching OFF circuit.

dv/dt Triggering

In reverse biased junction J2 acquires the characteristic like capacitor because of presence of charge across the junction, means junction J2 behaves like a capacitance. If the forward voltage is applied suddenly, a charging current through the junction capacitance Cj lead to turn ON the SCR.

The charging current i_c is given by:

$i_c = dQ/dt = d(Cj*Va) / dt$

$i_c = (Cj * dVa /dt) + (Va* dCj / dt)$

(where, Va is forward voltage appears across junction J2)

As the junction capacitance is nearly constant, dCj / dt is zero, then:

$i_c = Cj\ dVa / dt$

Therefore, if the rate of rise of forward voltage dVa /dt is high, the charging current i_c would be more. Here, the charging current plays the role of gate current to turn On the SCR even the gate signal is zero.

Temperature Triggering

When the Thyristor is in forward blocking mode, most of the applied voltage collects over the junc-tion J2, this voltage associated with some leakage current. Which increases the temperature of the junction J2. So, with the increase in temperature the depletion layer decrease and at some high temperature (within the safe limit), depletion layer breaks and the SCR turns to ON state.

Light Triggering

For triggering a SCR with light, a recess (or hollow) is made inner p-layer as shown in figure be-low. The beam of light of particular wavelength is directed by optical fibres for irradiation. As, the intensity of the light exceeds to a certain value, SCR get turn ON. These type of SCR called as Light Activated SCR (LASCR). Sometimes, these SCR triggered using both light source and gate signal in combination. High gate current and lower light intensity required to turn ON the SCR.

LASCR or Light triggered SCR are used in HVDC (High Voltage Direct Current) transmission system.

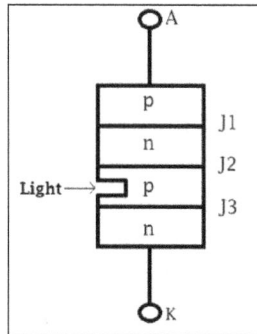

Difference between Thyristor and MOSFET

Thyristor and MOSFET both are electrical switches and are most commonly used. The basic difference between both of them is that MOSFET switches are voltage controlled device and can only switch DC current while Thyristors switches are current controlled device and can switch both DC and AC current.

There are some more differences between Thyristor and MOSFET are given below in table:

Property	Thyristor	MOSFET
Thermal Run away	Yes	No
Temperature sensitivity	less	high
Type	High voltage high current device	High voltage medium current device
Turning off	Separate switching circuit is required	Not required
Turning On	Single pulse required	No continuous supply is required except during turning On and Off
Switching speed	low	high
Resistive input impedance	low	high
Controlling	Current controlled device	Voltage controlled device

Difference between Thyristor and Transistor

Thyristor and Transistor both are electrical switches but the power handling capacity of Thyristors are is far better than transistor. Due to having high rating of Thyristor, given in kilowatts, while of transistor power ranges in watts. A Thyristor is taken as a closed couple pair of transistors in analysis. The main difference between the transistor and Thyristor is, Transistor need continuous switching supply to remain ON but in case of Thyristor we need to trigger it once only and it remains ON. For applications like alarm circuit which need to trigger once and stay ON forever, cannot use transistor. So, to overcome these problems we use Thyristor.

There are some more differences between Thyristor and Transistor is given below in table:

Property	Thyristor	Transistor
Layer	Four Layers	Three Layers
Terminals	Anode, Cathode and Gate	Emitter, Collector, and Base
Operation over voltage and current	Higher	Lower than thyristor
Turning ON	Just required a gate pulse to turn ON	Required continuous supply of the controlling current
Internal power loss	Lower than transistor	Higher

References

- Semiconductor-devices-types-and-applications: elprocus.com, Retrieved 15 February, 2019

- Diode-working-principle-and-types-of-diode: electrical4u.com, Retrieved 25 July, 2019

- Transistor: circuitglobe.com, Retrieved 6 January, 2019

- Transistor: electronics-tutorials.ws, Retrieved 16 March, 2019

- Npn-transistor: electronicshub.org, Retrieved 26 May, 2019

- Pnp-transistor: electronicshub.org, Retrieved 7 February, 2019

- Field-effect-transsitor-definition-and-types, analog-electronics: ecetutorials.com, Retrieved 17 January, 2019

- Mosfet-metal-oxide-semiconductor-basics, fet-field-effect-transistor, electronic-components: electronics-notes.com, Retrieved 27 March, 2019

- Applications-of-mosfet: electrical4u.com, Retrieved 8 June, 2019

- What-is-a-thyristor, electronic-components: electronics-notes.com, Retrieved 18 August, 2019

- What-is-thyristor-how-it-works: circuitdigest.com, Retrieved 28 April, 2019

Power Converters

An electrical circuit which changes the electric energy from one form into the required form optimized for the specific load is known as a power converter. There are numerous types of power converters such as analog to digital converter, digital-to-analog converter and inverter. The topics elaborated in this chapter will help in gaining a better perspective about these types of power converters.

A power converter is an electrical circuit that changes the electric energy from one form into the desired form optimized for the specific load.

The task of a power converter is to process and control the flow of electric energy by supplying voltages and currents in a form that is optimally suited for the user loads.

Energy was initially converted in electromechanical converters (mostly rotating machines). Today, with the development and the mass production of power semiconductors, static power converters find applications in numerous domains and especially in particle accelerators. They are smaller and lighter and their static and dynamic performances are better.

A static converter is a meshed network of electrical components that acts as a linking, adapting or transforming stage between two sources, generally between a generator and a load.

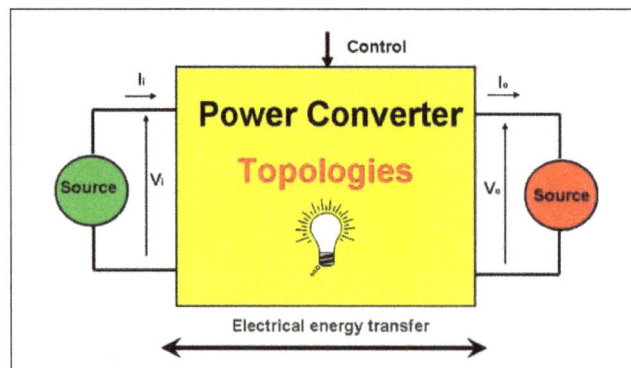

Power converter definition

An ideal static converter controls the flow of power between the two sources with 100% efficiency. Power converter design aims at improving the efficiency. But in a first approach and to define basic topologies, it is interesting to assume that no loss occurs in the converter process of a power converter. With this hypothesis, the basic elements are of two types:

- Non-linear elements, mainly electronic switches: semiconductors used in commutation mode.

- Linear reactive elements: capacitors, inductances and mutual inductances or transformers. These reactive components are used for intermediate energy storage but also for voltage and current filtering. They generally represent an important part of the size, weight, and cost of the equipment.

This introductory paper reviews and gives a precise definition of basic concepts essential for the understanding and the design of power converter topologies. First of all the sources and the switches are defined. Then, the fundamental connection rules between these basic elements are reviewed. From there, converter topologies are derived. Some examples of topology synthesis are given.

Finally, the concept of hard and soft commutation is introduced.

Sources

A power converter processes the flow of energy between two sources. To synthesize a power converter topology, the first step is to characterize these sources.

In the energy conversion process, a source is mainly a generator (called often input source) or a load (called output source). However, in the case of a change of direction of the energy flow, i.e., a change in the sign of the power, the sources (generators and loads) can exchange their functions (i.e., restitution of energy of a magnet towards the grid).

Nature of Sources

There are two types of sources: voltage and current sources. As mentioned earlier, any of these sources could be a generator or a receptor (load).

A source is called a voltage source if it is able to impose a voltage regardless of the current flowing through it. This implies that the series impedance of the source is zero (or negligible in comparison with the load impedance).

A source is called a current source if it is able to impose a current regardless of the voltage at its terminals. This implies that the series impedance of the source is infinite (or very large in comparison with the load impedance).

These definitions correspond to permanent properties. The principle of operation of a converter is based on the switch mode action of its switches. Commutations of the switches generate very fast current and voltage transients so that the transient behaviour of the sources is fundamental for converter design. The transient behaviour of a source is characterized by its ability or inability to withstand steps generated by the external circuit in the voltage across its terminals or in the current flowing through it. Then new definitions could be stated:

- A source is a voltage source if the voltage across its terminals cannot undergo a discontinuity due to the external circuit variation. The most representative example is the capacitor since an instantaneous change of voltage across its terminals would mean an instantaneous change of its charge which would require an infinite current.

- A source is a current source if the current flowing through it cannot undergo a discontinuity due to the external circuit variation. The most representative example is the inductance since an instantaneous change in current would correspond to an instantaneous change in its flux which would require an infinite voltage.

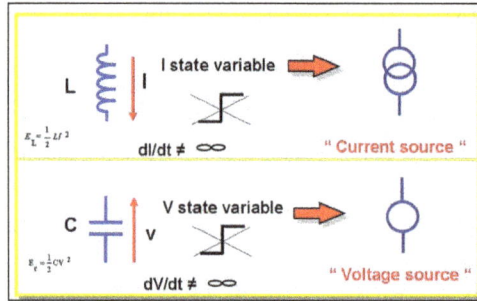

Inductance and capacitor versus current and voltage source

It should be noted that a square wave voltage generator (respectively a current generator) is indeed a voltage source (respectively a current source) as defined above since the voltage steps (respectively current steps) are not caused by the external circuit.

With these definitions, it is interesting to define the notion of instantaneous impedance of a source as the limit of the source impedance when the Laplace operator tends towards infinity. Theoretically this instantaneous impedance can be zero, finite or infinite.

A source is referred to as a voltage source when its instantaneous impedance is zero, while a source is called a current source if its instantaneous impedance is infinite.

For example:

- Capacitor: $Z(s) = 1/(Cs)$, $\lim_{s \to \infty} Z(s) = 0 \Rightarrow$ voltage source.

- Inductance: $Z(s) = Ls$, $\lim_{s \to \infty} Z(s) = \infty \Rightarrow$ current source.

The determination of the source reversibilities is fundamental. The static characteristics of the switches can be derived from the reversibility analysis.

The voltage (or the current) that characterizes a source is called DC if it is unidirectional. As a first approximation, it can be taken as constant. The voltage (or the current) is called AC if it is periodic and has an average value equal to zero. As a first approximation, it can be taken as sinusoidal.

A source is voltage-reversible if the voltage across its terminals can change sign. In the same way, a source is current-reversible if the current flowing through it can reverse.

In summary, the input/output of a converter can be characterized as voltage or current sources (generator or loads), either DC or AC, current-reversible and/or voltage-reversible. There are only eight possibilities, as shown in figure.

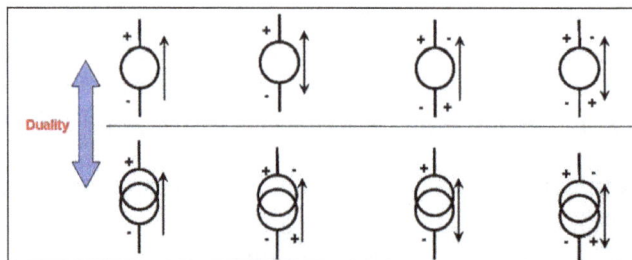

Voltage and current sources with their reversibilities

Source Nature Modification

Connecting in series an inductance with an appropriate value to a voltage source (that is a dipole with zero instantaneous impedance) turns the voltage source into a current source. In the same way, connecting in parallel a capacitor of appropriate value to a current source (dipole with infinite instantaneous impedance) turns the current source into a voltage source.

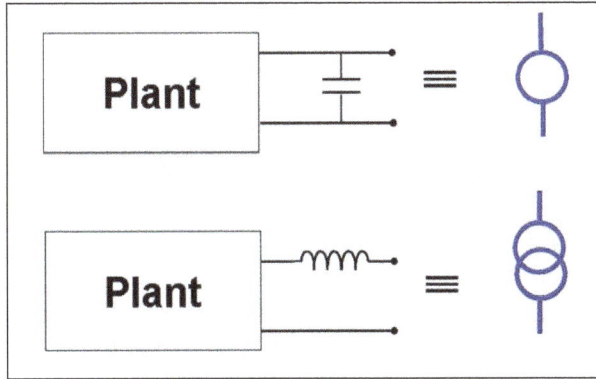

Source nature confirmation or modification

These inductive or capacitive elements connected in parallel or in series with the source can temporarily store energy. Consequently, if an inductance connected to a voltage source turns it into a current source, it is important to determine the current reversibility of this source.

In practice, the identification of a real generator or of a real load with a voltage or current source is not obvious. This is why the nature of the source is often reinforced by the addition of a parallel capacitor in the case of voltage sources and by the addition of a series inductor in the case of current sources.

The current source obtained by connecting an inductance in series with a voltage source keeps the same current reversibility as the voltage source. The inductance acts as a buffer absorbing the voltage differences. Consequently, the current source obtained by connecting a series inductance to a voltage source is reversible in voltage. When the voltage source itself is reversible in voltage there is no particular problem. But, if the voltage source is not reversible in voltage, the current source obtained by connecting a series inductance to the voltage source is only instantaneously reversible with respect to voltage.

The former result can easily be transposed to the voltage source obtained by connecting a capacitor in parallel with a current source. The voltage source thus obtained keeps the same voltage reversibility as the current source and is reversible in current. However, this reversibility is only instantaneous if the current source is not reversible in current.

Example: The set of ideal batteries shown in figure. behaves as a load during charging and as a generator during discharging; such a source is called a DC voltage source, current-reversible but not voltage-reversible. Nevertheless, because of the inductance of the connecting cables, this battery can sometimes be taken as a current source instantaneously voltage-reversible and permanently current-reversible. If a capacitor bank is added at the terminals of the cables, it becomes again a voltage source.

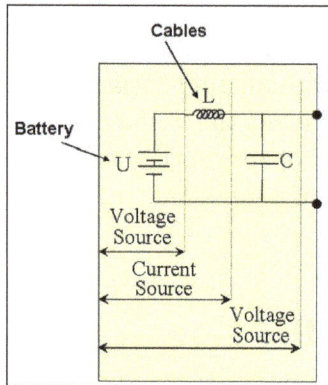

Source nature modification and confirmation

Switches Characteristics

Static converters are electrical networks mainly composed of semiconductor devices operating in the switch mode (switches); through proper sequential operation of these components, they transfer energy between two sources with different electrical properties.

Minimizing the losses in the switches maximizes the efficiency of the converter. These switches must have a voltage drop (or an ON resistance) as low as possible in the ON-state, and a negligible leakage current (or an OFF resistance) in the OFF-state. These two states are defined as static states.

The change from one state to the other (switch commutation) implies a transient behaviour of the switch. This behaviour is complex because it depends on the control of the switch (through a gate control) and on the conditions imposed by the external circuit.

Static Characteristics

In the static domain, a switch has the same behaviour as a non-linear resistance: very low in the ON state and very high in the OFF-state.

Taken as a dipole with the load sign convention, the static characteristics $I_k(V_k)$, which represents the operating points of a switch, are made up of two branches totally located in quadrants 1 and 3 such that $V_k \times I_k > 0$. One of these branches is very close to the I_k axis (ON-state) and the other is very close to the V_k axis (OFF-state). Each of these branches can be located in one or two quadrants. In the case of an ideal switch, the static characteristics are the half-axis to which they are close.

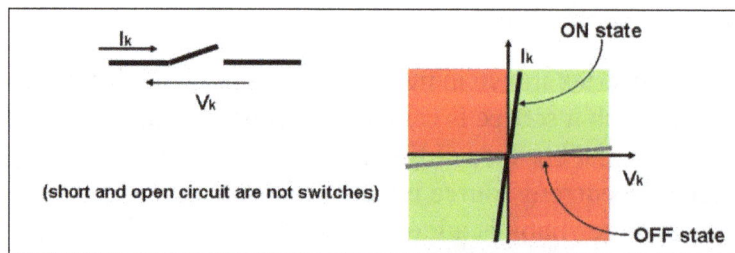

Static characteristics of a switch

In this representation, any switch which behaves really as a switch (commutation: ON <=> OFF), has a static characteristic consisting of at least two orthogonal half-axes (or segments) except for the obvious cases of the short circuit and of the open circuit which correspond, respectively, to a switch always ON and to a switch always OFF.

The static characteristics, an intrinsic feature of a switch, are reduced to a certain number of segments in the $I_k(V_k)$ plane:

- Two-segment characteristics: The switch is unidirectional in current and in voltage. Two two-segment characteristics can be distinguished: in the first case, current I_k and voltage V_k are of the same sign; in the second case, current I_k and voltage V_k are of opposite sign. The switches having such characteristics are respectively called T and D switches.

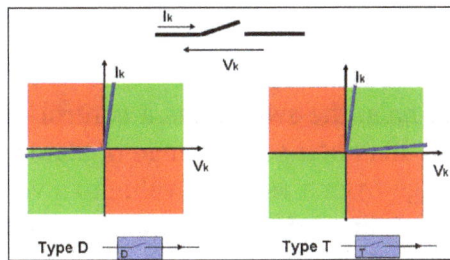

Static characteristics of a two-segment switch

- Three-segment characteristics: The switch is bidirectional in current and unidirectional in voltage, or vice versa. Therefore there are two types of three-segment static characteristics. Note that these two types of switches could be synthesized with the association in parallel or in series of two-segment switches (T and D).

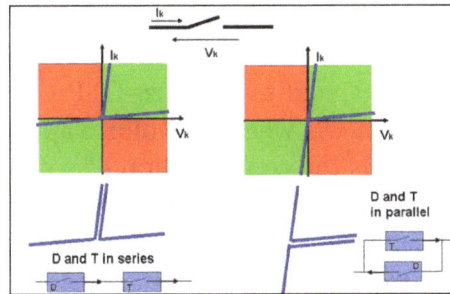

Static characteristics of a three-segment switch

- Four-segment characteristics: The switch is bidirectional in voltage and in current. There is only one such type of static characteristic. A four-segment characteristic could be obtained by a connection in series or in parallel of switches with three-segment characteristics.

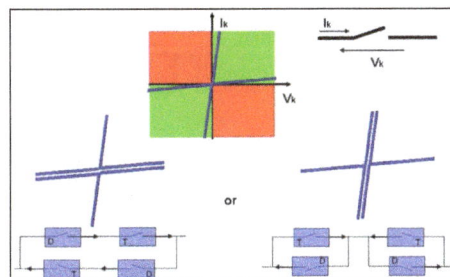

Static characteristics of a four-segment switch

Dynamic Characteristics

The dynamic characteristics is the trajectory described by the point of operation of the switch during its commutation, going from one half-axis to the perpendicular half-axis. A switch being either ON or OFF, there are two commutation dynamic characteristics corresponding to the turn-ON and to the turn-OFF, which will be grouped under the global term of dynamic characteristics.

Unlike the static characteristics, the dynamic characteristics is not an intrinsic property of the switch but depends also on the constraints imposed by the external circuit. Neglecting second-order phenomena and taking into account the dissipative nature of the switch, the dynamic characteristics can only be located in those quadrants where $V_k \times I_k > 0$ (generator quadrants). For the two commutations (turn-ON and turn-OFF), two modes are possible: controlled commutation and spontaneous commutation.

Controlled Commutation

In addition to its two main terminals, the switch has a control terminal on which it is possible to act in order to provoke a quasi-instantaneous change of state (case of T switch). The internal resistance of this switch can change from a very low value to a very high value at turn-OFF (and inversely at turnON). These changes are independent of the evolution of the electrical quantities imposed on the switch by the external circuit.

It should be noted that in a controlled commutation, the switch imposes its state on the external circuit. Under such circumstances, the element can undergo severe stresses that depend on its dynamic characteristics. If the switching time is long and the operating frequency is high, the commutation losses can be important.

Spontaneous Commutation

The spontaneous commutations correspond to the turn-OFF when the current flowing through the switch reaches zero and to the turn-ON when the voltage applied across its terminals reaches zero. The spontaneous commutation is the commutation of a simple PN junction (D switch). It only depends on the evolution of the electrical variables in the external circuit. Spontaneous commutations could be achieved with any controlled semiconductor if the gate control is synchronized with the electrical variables of the external circuit. A spontaneous commutation is achieved with minimum losses since the operating point moves along the axes.

It is important to point out that a controlled commutation can happen only in the first or third quadrants while a spontaneous commutation can happen only with a change of quadrant.

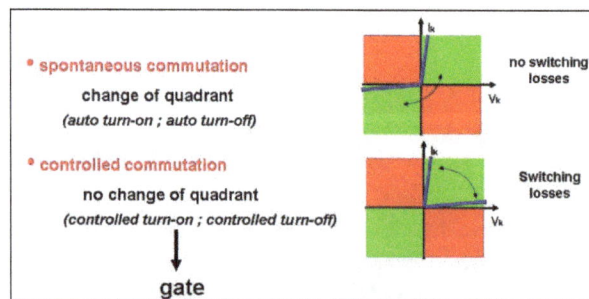

Spontaneous and controlled commutations

Classification of Switches

Finally, switches used in power converters can be classified by their static characteristics (two, three or four segments) and by the type of commutation (controlled or spontaneous) at turn-ON and at turn-OFF.

Two-segment Switches

Two switches with two-segment characteristics can be distinguished (not counting the open circuit and the short circuit):

- The first of these switches has the static characteristics of a D switch and its turn-ON and turnOFF commutations are spontaneous. This switch is typically a diode.

- The second of these switches has the static characteristics of a T switch and its turn-ON and turn-OFF commutations are controlled. It is a power semiconductor: MOSFET, IGBT, GTO, IGCT, etc.

This switch will be symbolized by separating the turn-ON and turn-OFF control gates as shown on figure.

A switch with two-segment characteristics similar to the T type (both segments in the same quadrant), must have controlled turn-ON and turn-OFF commutations. Should it have only one controlled commutation, it would then be necessary to put in series or in parallel a D switch (a diode) to get the spontaneous commutation. In this case, it is no longer a two-segment switch but a three-segment one. Therefore, only two two-segment switches can be used directly.

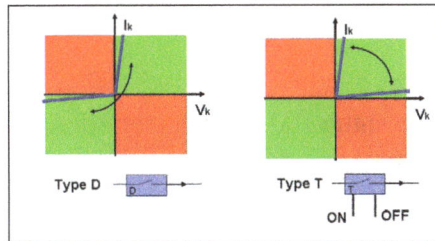

Dynamic characteristics of two-segment switches

Three-segment Switches

These switches can be divided into two groups depending on whether they are:

- Unidirectional in current and bidirectional in voltage.

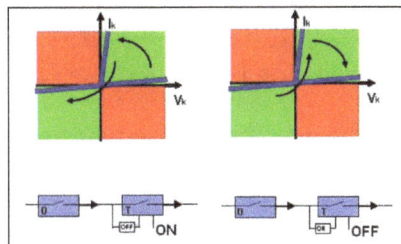

Dynamic characteristics of three-segment switches: bidirectional in voltage

- Bidirectional in current and unidirectional in voltage.

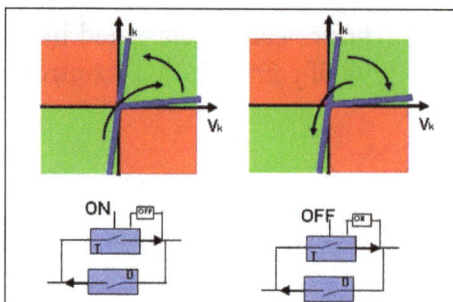

Dynamic characteristics of three-segment switches: bidirectional in current

Except for the thyristor, all these switches are synthesized switches, realized by connecting a diode in parallel or in series with a two-segment switch. A special driver is needed to obtain the spontaneous commutation. The 'dual-thyristor' (unidirectional in voltage, bidirectional in current, controlled turn-OFF and spontaneous turn-ON) is a good example of a useful three-segment switch.

In each of these two groups, all switches have the same static characteristics but differ by their commutation mechanisms. It is important to remark that if a three-segment switch has both commutations controlled (turn-ON and turn-OFF) or both spontaneous, it would never use the three segments of its static characteristics. Therefore, a three-segment switch must necessarily have one controlled commutation and one spontaneous commutation.

The cycle of operation, which represents the locus described by the point of operation of these switches, is then fully determined. They can only be used in converter topologies which impose a single cycle of operation on the switches.

Four-segment Switches

All four-segment switches have the same static characteristics. They differ only by their commutation modes that can, a priori, differ in quadrants 1 and 3. So, six four-segment switches can be distinguished.

These switches are used mainly in direct frequency changers and in matrix converters; in practice they are made up of two three-segment switches connected in series or in parallel.

Interconnection of Sources

Commutation Rules

To control the power flow between two sources, the principle of operation of a static power converter is based on the control of switches (turn-ON and turn-OFF) with particular cycles creating periodic modifications of the interconnection between these two sources.

The source interconnection laws can be expressed in a very simple way:

- A voltage source should never be short-circuited but it can be open-circuited.

- A current source should never be open-circuited but it can be short-circuited.

From these two general laws, it can be deduced that a direct connection between two voltage sources or between two current sources cannot be established by means of switches, as shown in figure.

Basic interdictions of source interconnection

In the case of two voltage sources, the switch turn-ON can only happen when the two sources have the same values, that is to say at the zero crossing of the voltage across the switch. The turn-ON must then be spontaneous (since it depends on the external circuit) and the turn-OFF can be controlled at any time. In the case of two current sources, the switch turn-OFF can only happen when the two current sources reach the same value, that is to say when the current in the switch reaches zero. In this case, the turn-OFF is spontaneous and the turn-ON can be controlled at any time.

Thus it is obvious that capacitors can be connected in parallel and inductances in series but with zero voltage, respectively, zero current, as is routinely done.

The previous laws forbid the commutation of switches between two sources of the same nature.

Structure of Power Converters

A power converter can be designed with different topologies and with one or several intermediate conversion stages. When this conversion is achieved without an intermediate energy storage stage, the conversion is called direct conversion and is achieved by a direct converter. On the other hand, when this conversion makes use of one or more stages storing energy temporarily, the conversion is called indirect and is achieved by an indirect converter.

The prohibition on connecting two sources of the same nature gives relevance to the following two classes of basic conversion topologies:

- Direct link topology: when the two sources have different natures.

- Indirect link topology: when the two sources have the same nature.

Direct Link Topology Converters

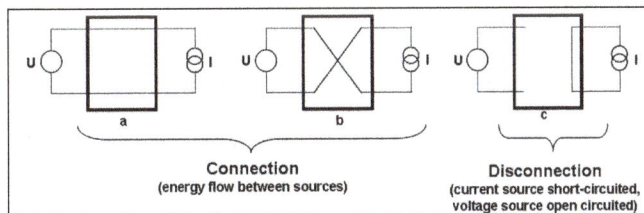

Interconnection possibilities between a voltage and a current source

A direct converter is an electrical network composed of switches only and is unable to store energy. In such a converter, the energy is directly transferred from the input to the output (assuming the losses can be neglected); the input power is equal to the output power at any time.

Taking into account the interconnection rules recalled above, the possible connections between a voltage source and a current source are shown in figure. The simplest structure allowing all these connections is the four-switches bridge:

- With K1 and K3 closed, the connection of above figure (a) is obtained.

- With K2 and K4 closed, the connection of above figure (b) is obtained.

- With K1 and K2 closed (or with K3 and K4 closed), the connection of above figure (c) is obtained.

When some of these connections are not necessary, it is possible to replace the bridge structure by simpler structures using fewer switches (i.e. buck converter).

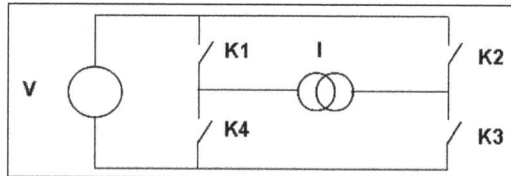

Basic configuration of voltage–current direct converter

The energy conversion between an input current source and an output voltage source poses the same problem. The basic configuration is the same but it is more usual to represent the input source on the left and the output source on the right.

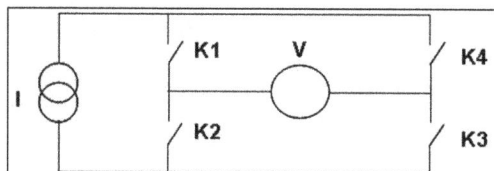

Basic configuration of current–voltage direct converter

Indirect Converters

It is not possible to directly interconnect two sources of the same nature with switches. It is necessary to add components to generate an intermediate buffer stage of a different type without active energy consumption (capacitor or inductance). This buffer stage is a voltage source (capacitor) if the energy transfer is between two current sources, and it is a current source (inductance) if the energy transfer is between two voltage sources.

Modification of the Nature of the Input or Output Source

In the case of the voltage–voltage conversion, one solution consists in adding an inductance in series with the input voltage source or with the output voltage source. With this change of the nature of the source, it is possible to use a direct converter: current–voltage or voltage–current converter according to where the inductance was added.

The case of the current–current conversion is the opposite of the previous case. A capacitor is added in parallel or in series with the input or with the output current source.

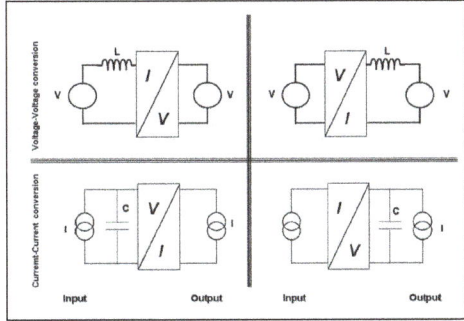

Indirect converters: modification of the input or output source nature

Use of two Direct Converters

If it is not possible or too costly to modify the nature of one source, two direct converters can be connected with an intermediate buffer stage: an inductance for a voltage–voltage conversion and a capacitor for a current–current conversion.

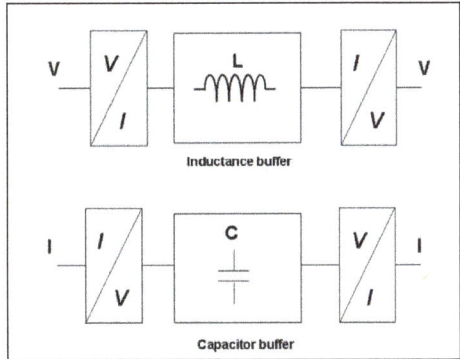

Indirect converters: intermediate buffer stage and two direct converters

Voltage–voltage Indirect Converters

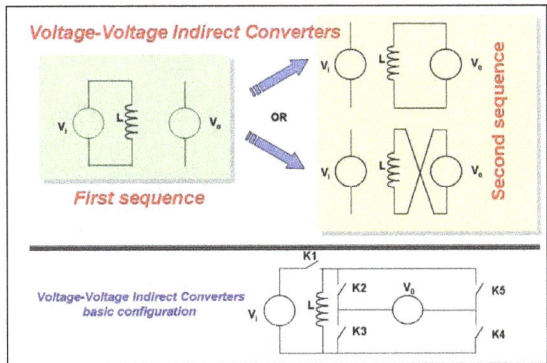

Voltage–voltage indirect converters

In the indirect converters, the two voltage sources are never connected directly. Two sequences are then necessary. During the first sequence, the energy is transferred from the input voltage source to the inductance (voltage to current conversion). During the second sequence, the inductance gives back the energy to the output voltage source (current to voltage conversion; two directions are possible). An extra switch is necessary to get these sequences.

Current–current Indirect Converters

In the indirect converters, the two current sources are never connected directly. Two sequences are then necessary. During the first sequence, the energy is transferred from the input current source to the capacitor (current to voltage conversion). During the second sequence, the capacitor gives back the energy to the output current source (voltage to current conversion; two directions are possible).

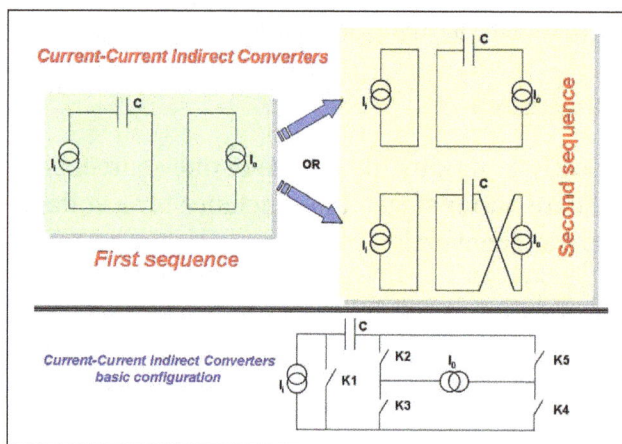

Current–current indirect converters

Figure represents the three basic configurations of all single-phase converters.

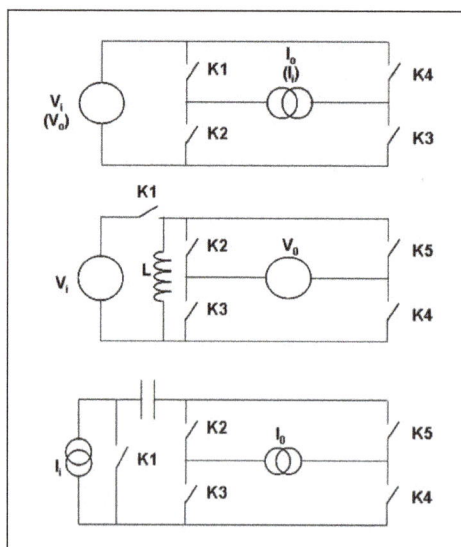

Three basic structures of single-phase power converters

From these basic configurations, it is possible, according to the nature of the sources, to associate them or to add other components. For example, in the case where one of the sources is AC, it is possible to insert a transformer for adaptation or galvanic insulation.

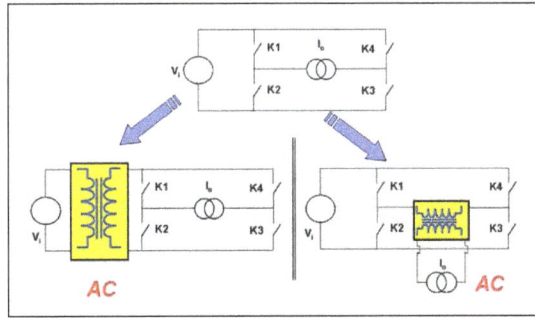

Insertion of a transformer in a direct topology

It is also possible to associate several basic topologies. One of the more usual applications is to have an intermediate stage at high frequency to reduce the size of the transformers and magnetic elements and to get a higher performance at the output (bandwidth, ripple, etc.)

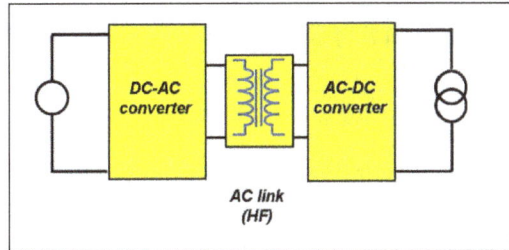

Association of elementary structures

Figure illustrates the case of resonant converters. Note that the interconnection of the various intermediate sources must respect the interconnection laws. It is especially important when choosing the output filter.

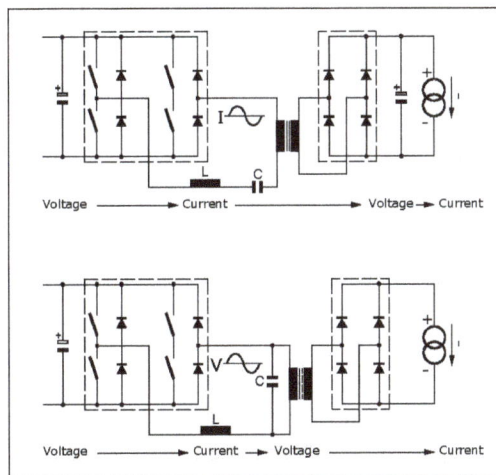

Series and parallel resonant converters

Power Converter Classification

Figure summarizes in tabular form the power conversion topologies:

- The crossed cells correspond to reversibility incompatibilities between the input and output sources.

- Two symmetric cells, with respect to the diagonal D, represent two reversible topologies.

- Two topologies corresponding to two cells symmetric with respect to the point O are dual.

Analog to Digital Converter

Analog-to-Digital converters (ADC) translate analog signals, real world signals like temperature, pressure, voltage, current, distance, or light intensity, into a digital representation of that signal. This digital representation can then be processed, manipulated, computed, transmitted or stored.

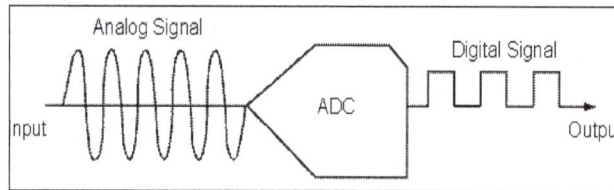

Analog to Digital conversion

In many cases, the analog to digital conversion process is just one step within a larger measurement and control loop where digitized data is processed and then reconverted back to analog signals to drive external transducers. These transducers can include things like motors, heaters and acoustic divers like loudspeakers. The performance required of the ADC will reflect the performance goals of the measurement and control loop. ADC performance needs will also reflect the capabilities and requirements of the other signal processing elements in the loop.

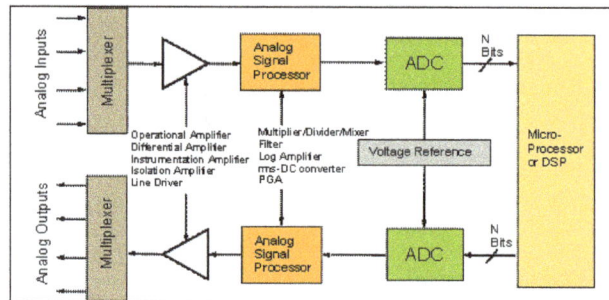

Figure Measurement and Control Loop

Basic Operation

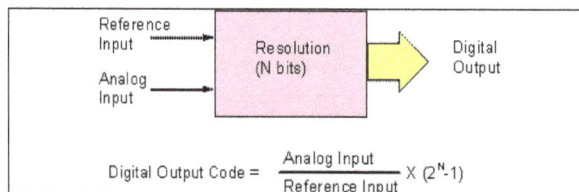

Digital output code

An ADC samples an analog waveform at uniform time intervals and assigns a digital value to each sample. The digital value appears on the converter's output in a binary coded format. The value is obtained by dividing the sampled analog input voltage by the reference voltage and them multiplying by the number of digital codes. The resolution of converter is set by the number of binary bits in the output code.

An ADC carries out two processes, sampling and quantization. The ADC represents an analog signal, which has infinite resolution, as a digital code that has finite resolution. The ADC produces 2N digital values where N represents the number of binary output bits. The analog input signal will fall between the quantization levels because the converter has finite resolution resulting in an inherent uncertainty or quantization error. That error determines the maximum dynamic range of the converter.

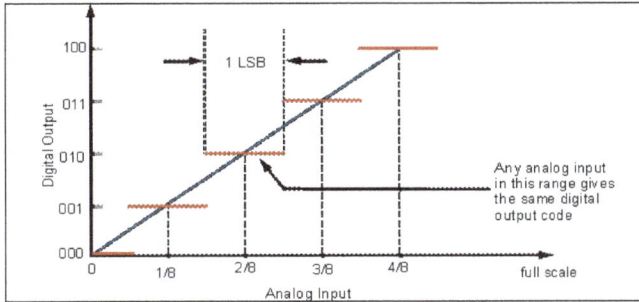

Quantization Process

The sampling process represents a continuous time domain signal with values measured at discrete and uniform time intervals. This process determines the maximum bandwidth of the sampled signal in accordance with the Nyquist Theory. This theory states that the signal frequency must be less than or equal to one half the sampling frequency to prevent aliasing. Aliasing is a condition in which frequency signals outside the desired signal band will, through the sampling process, appear within the bandwidth of interest. However, this aliasing process can be exploited in communications systems design to down-convert a high frequency signal to a lower frequency. This technique is known as under-sampling. A criterion for under-sampling is that the ADC has sufficient input bandwidth and dynamic range to acquire the highest frequency signal of interest.

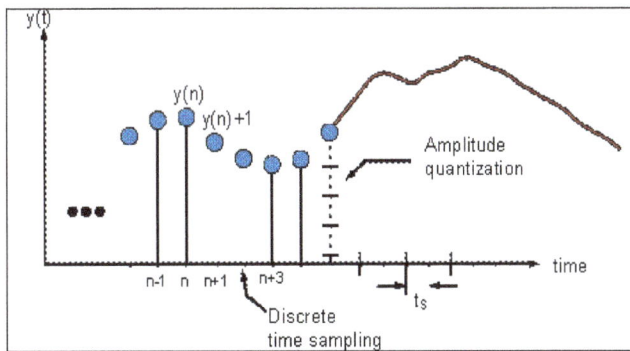

Sampling Process

Sampling and quantization are important concepts because they establish the performance limits of an ideal ADC. In an ideal ADC, the code transitions are exactly 1 least significant bit (LSB) apart. So, for an N-bit ADC, there are 2N codes and 1 LSB = FS/2N, where FS is the full-scale analog input voltage. However, ADC operation in the real world is also affected by non-ideal effects, which produce errors beyond those dictated by converter resolution and sample rate. These errors are reflected in a number of AC and DC performance specifications associated with ADCs.

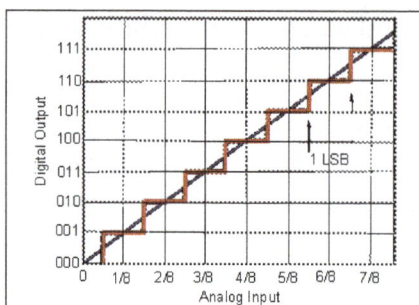

Transfer function for an ideal ADC

Any analog input in this range gives the same digital output code.

Understanding Key Specifications

Specification and terms, units of measure	Meaning	Significance
DC specifications		
Resolution or bits	Number of bits representing an analog signal, generally ranging from 6 to 24.	Determines how small an input can be resolved.
Conversion speed or rate, ksamples/s or Msamples/s	The number of repetitive conversions per second for a full-scale change to specified resolution and linearity.	Determines the fastest sampling capability of the ADC.
Least significant bit (LSB)	The right-most bit in an ADC output code. LSB size is a function of converter resolution.	Not a specification, but a common term.
Most significant bit (MSB)	The left-most bit in an ADC output code.	Not a specification, but a common term.
Differential nonlinearity (DNL), expressed in terms of LSB	The deviation from the ideal (1 LSB) code width between any two adjacent codes. In an ideal converter, every code is exactly the same size and DNL is zero.	DNL, INL, offset error, and gain error specify how accurately the data represents the signal across the entire internal and external range.
Integral nonlinearity (INL), expressed in terms of LSB (also referred to as "relative accuracy error")	The deviation of an actual code transition point from its ideal position on a straight line drawn between the end points of the transfer function.	The narrowing or widening of code widths caused by DNL can lead to "missing codes" and add noise and frequency spurs beyond the effects of quantization.
Offset, expressed in terms of LSB	The difference between the ideal and actual output when the converter input is zero.	INL describes the absolute accuracy of a converter. Calculated after offset and gain errors are removed.
Gain error/full-scale error, expressed in terms of LSB	The difference between the ideal and actual output when the converter input is at full scale.	INL produces additional harmonics and spurs in the frequency domain.

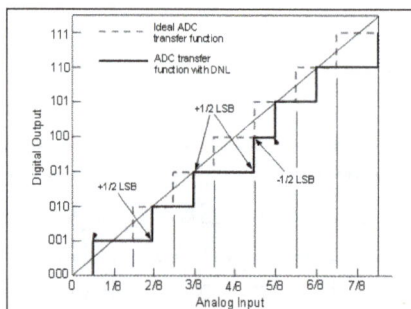

ADC transfer function with DNL error

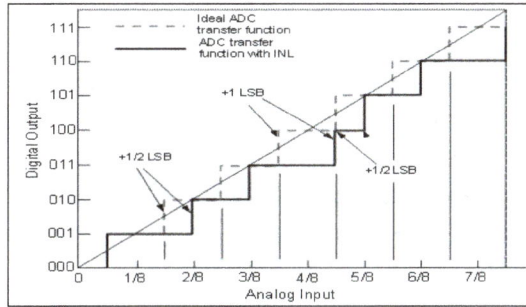

ADC transfer function with INL error

Specification and terms, units of measure	Meaning	Significance
Spurious-free dynamic range (SFDR), dB applications where a spur may interfere with a neighboring channel.	The ratio of the fundamental frequency's amplitude to that of the largest spurious signal in a given bandwidth.	Important in communications.
Total harmonic distortion (THD), dB	The ratio of the rms sum of the first six harmonics to the amplitude of the fundamental frequency.	Harmonics are noise components related to, or generated by, a-d conversion. Harmonics can limit the dynamic performance of a converter.
Signal-to-noise-and-distortion ratio (SINAD), dB	The ratio of the rms signal amplitude to the mean value of the root-sum-squares (RSS) of all other spectral components including harmonics but excluding dc.	SINAD indicates the true ac linearity an ADC because it includes the effects of the 2nd and 3rd harmonics.
Effective number of bits (ENOB) performance of a given ADC as compared to an ideal converter.	ENOB = SINAD ...1.76 dB 6.02	ENOB specifies the dynamic.
Signal-to-noise ratio (SNR) or signal-to-noise ratio without harmonics.	Similar to SINAD, the ratio of the rms signal amplitude to the mean value of the root-sum-squares of all other spectral components, excluding the first five harmonics and dc.	SNR indicates noise performance of a converter compared to an ideal converter.
Analog bandwidth (full-power, small signal), kHzor MHz	The input frequency where the fundamental in an FFT of the output rolls off by 3 dB. Generally determined by the Converter's sample-and-hold amplifier.	Important in IF under-sampling applications. This spec may not be compatible with the ADC's maximum sampling rate.
Power dissipation, mW or W	The amount of power consumed by the converter.	Important for power-sensitive applications in which battery life, temperature, or space limitations may affect power dissipation requirements.

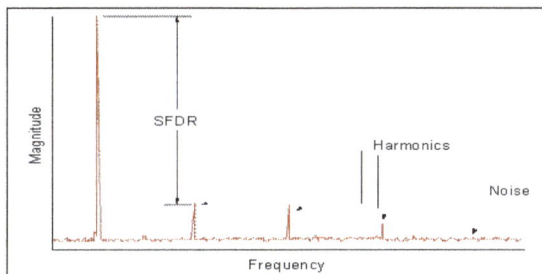

Frequency domain specifications

ADC Classifications

Speed and accuracy are two critical measures of ADC performance. As such, they provide a means for broadly categorizing today's monolithic ADCs. ADC chips may be loosely grouped along these lines as general-purpose, high-speed, or precision. Converters with 8- to 14-bit resolution and conversion rates below 10 Msamples/s are typically considered general-purpose ADCs. Those with conversion rates above 10 Msamples/s usually get the high-speed moniker, while those with 16 bits or more of resolution fall into the precision ADC category.

These definitions, however, are somewhat arbitrary and largely reflect the current state-of-the-art.

Within these broad categories, ADCs may also be grouped according to converter architecture. The most popular types are flash, pipelined, successive approximation- register, and sigma-delta. Each architecture offers certain advantages with respect to conversion speed, accuracy, and other parameters. The characteristics associated with each architecture help determine its suitability for a given application.

ADCs have been implemented both as discrete Designs' sometimes constructed with hybrid packaging and as monolithic designs implemented as integrated circuits (ICs). Development of monolithic ADCs has been heavily influenced by process innovation, both in high-end processes such as bipolar, biCMOS, and SiGe, as well as mainstream CMOS processes.

Over time, the migration of ADC designs to CMOS processes with smaller geometries has increased the possibilities for performance enhancements, while also allowing higher levels of integration. That integration can increase the number of conversion channels achieved on a single die, or allow conversion- related functions to be brought on-chip. As a result, die size and, consequently, package size depends on the semiconductor process employed. The process also determines supply voltage, which along with conversion speed, influences power dissipation.

Flash Architecture

In the flash or parallel ADC architecture, an array of 2N-1 comparators converts an analog signal to digital with a resolution of N bits. The comparators receive the analog signal on one input and a unique fraction of the reference voltage on the other. The reference voltage for each comparator is often a tap off a resistive voltage divider, whereby the comparators are biased in voltage increments equivalent to 1 LSB. The comparator array is clocked simultaneously.

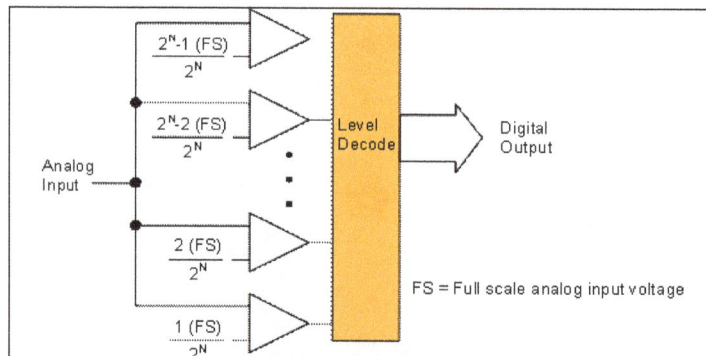

Flash Architecture

The comparators with reference voltages less than the analog input will output a digital one. The comparators with reference voltages greater than the analog input will output a digital zero. When read together, the outputs present a "thermometer code," which the output logic converts to standard binary code.

Pros:

- Very fast, converts in one ADC clock cycle.

Cons:

- Requires many comparators. The physical limits of monolithic integration generally allow only up to 8 bits of resolution per ADC chip.

- High input capacitance.

Pipeline Architecture

This architecture divides the conversion into two or more stages. Each stage consists of a sample-and-hold (S/H) circuit, an m-bit flash ADC, and a DAC. The analog signal is fed to the first stage, where it's sampled by the S/H and converted to a digital code by the flash ADC. The code generated by the flash ADC in this stage represents the most significant bits of the ADC's final output.

The same code is then fed to the DAC, which reconverts the code back to an analog signal that is subtracted from the original, sampled analog input signal. The resulting difference signal or residue, is next amplified and sent on to the following stage in the pipeline, where the whole process is repeated. The number of stages needed depends on the required resolution and the resolution of the flash ADCs used in each stage. In theory, the overall resolution of the ADC would be the sum of the resolutions of the flash ADCs. But in practice, some extra overlapping bits are required for error correction.

Pros:

- Not as fast as pure flash architecture, but achieves higher resolutions and dynamic range.

- Handles wideband inputs.

- Use of dither noise and averaging increases the effective resolution of the converter.

- Permits under sampling of wideband IF signal.

Cons:

- Pipeline delay. Total throughput can be equal to that of a flash converter (one conversion per cycle), but with a latency or pipeline delay equal to the number of stages.

- Accuracy of conversion depends on the DAC linearity.

- Ill-suited to applications where conversion results must be available immediately after the sample clock.

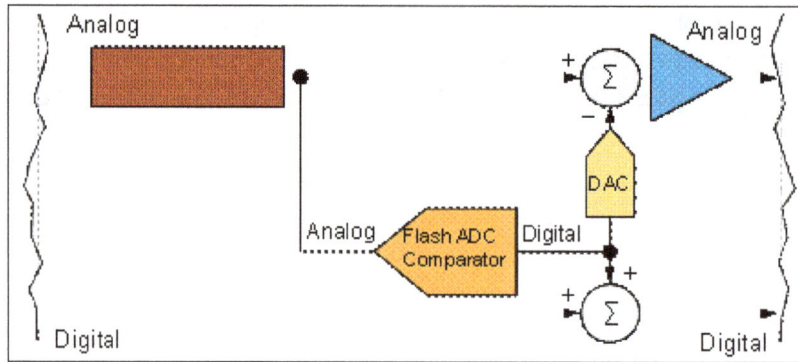

Single pipelined converter stage

SAR Architecture

The SAR converter works like a balance scale that compares an unknown weight against a series of known weights. In lieu of weights, the SAR converter compares the analog input voltage against a series of successively smaller voltages representing each of the bits in the digital output code. These voltages are fractions of the full-scale input voltage (1/2, 1/4, 1/8, 1/16...$1/2^N$, where N=number of bits).

The first comparison is made between the analog input voltage and a voltage representing the most significant bit (MSB). If that analog input voltage is greater than the MSB voltage, the value of the MSB is set to 1, otherwise it's set to 0. The second comparison is made between the analog input voltage and a voltage representing the sum of the MSB and the next most significant bit. The value of the second most significant bit is then set accordingly. The third comparison is made between the analog input voltage and the voltage representing the sum of the three most significant bits. At this point, the value of the third most significant bit is set. The process repeats until the value of the LSB is established.

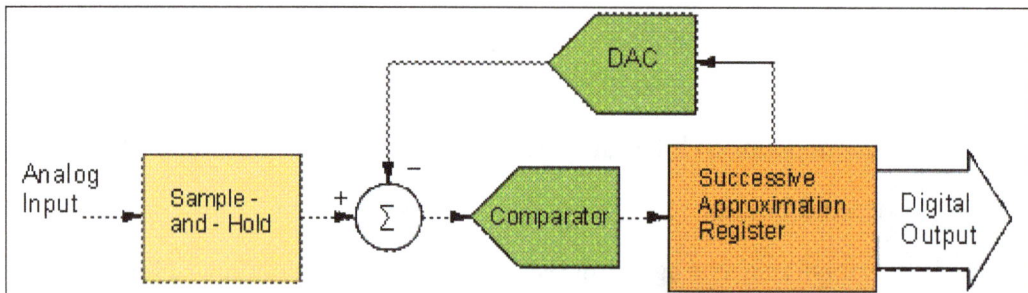

Successive-approximation-register (SAR) ADC

Pros:

- Uses a single comparator to achieve high resolution, resulting in small die size for monolithic ADCs.

- No pipeline delay.

- Well-suited for non-periodic inputs.

- Use of dither noise and averaging increases the effective resolution of the converter.

- Permits under-sampling.

Cons:

- Requires N comparisons to achieve N-bit resolution, which is more than both flash and pipelined.

- Accuracy of conversion depends on the DAC linearity and comparator noise.

Sigma-delta Architecture

The basic elements of this architecture are an integrator, a comparator, and a one-bit DAC, which together form a sigma-delta modulator. The modulator subtracts the DAC from the analog input signal and then feeds the signal to the integrator. The output of the integrator then goes to a comparator, which converts the signal to a one-bit digital output. The resulting bit is fed to the DAC, which produces an analog signal to be subtracted from the input signal. The process repeats at a very fast "oversampled rate.

The modulator produces a binary stream in which the ratio of ones to zeros is a function of the input signal's amplitude. By digitally filtering and decimating this stream of one and zeroes, a binary output representing the value of the analog input is obtained.

Pros:

- Yields the highest precision for lower input-bandwidth applications.

- Permits noise shaping whereby low-frequency noise is moved to higher frequencies, outside the band of interest.

- Oversampling reduces requirements for anti-aliasing filtering.

Cons:

- Latency is much greater than with other architectures.

- Oversampling and latency discourage the use of sigma-delta ADCs when digitizing multiplexed input signals.

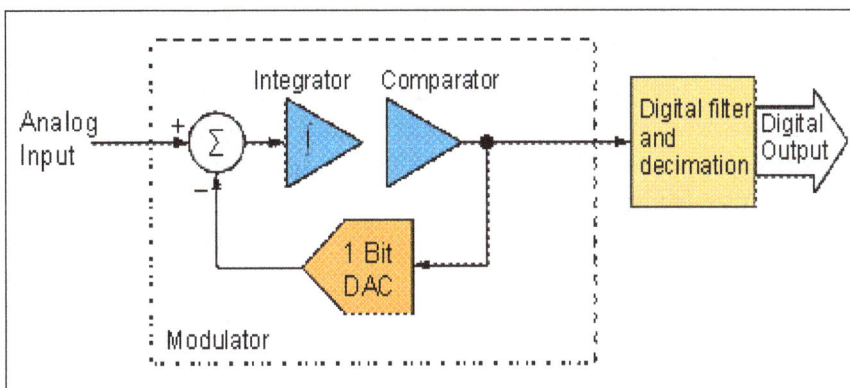

Sigma-delta architecture

Digital-to-Analog Converter

A digital-to-analog converter (DAC or D-to-A) is a device for converting a digital (usually binary) code to an analog signal (current, voltage or charges). Digital-to Analog Converters are the interface between the abstract digital world and the analog real life. Simple switches, a network of resistors, current sources or capacitors may implement this conversion.

A DAC inputs a binary number and outputs an analog voltage or current signal. In block diagram form, it looks like this:

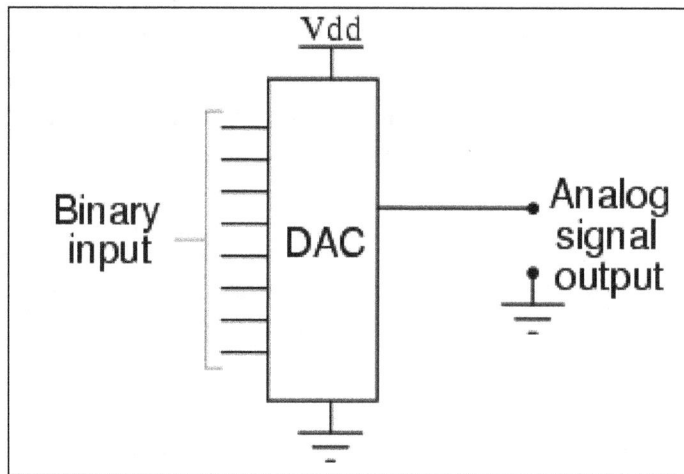

DAC Types

The most common types of electronic DAC's are:

- The Pulse Width Modulator the simplest DAC type. A stable current (electricity) or voltage is switched into a low pass analog filter with a duration determined by the digital input code. This technique is often used for electric motor speed control, and is now becoming common in high-fidelity audio.

- Oversampling DACs such as the Delta-Sigma DAC, a pulse density conversion technique. The oversampling technique allows for the use of a lower resolution DAC internally. A simple 1-bit DAC is often chosen, as it is inherently linear. The DAC is driven with a pulse density modulated signal, created through negative feedback. The negative feedback will act as a high pass filter for the quantization (signal processing) noise, thus pushing this noise out of the pass-band. Most very high resolution DACs (greater than 16 bits) are of this type due to its high linearity and low cost. Speeds of greater than 100 thousand samples per second and resolutions of 24 bits are attainable with Delta-Sigma DACs. Simple first order Delta-Sigma modulators or higher order topologies such as MASH - 'Multi stage' noise Shaping can be used to generate the pulse density signal. Higher oversampling rates relax the specifications of the output Low-pass filter and enable further suppression of quantization noise.

- The Binary Weighted DAC, which contains one resistor or current source for each bit of the DAC connected to a summing point. These precise voltages or currents sum to the correct output value. This is one of the fastest conversion methods but suffers from poor accuracy because of the high precision required for each individual voltage or current. Such high-precision resistors and current sources are expensive, so this type of converter is usually limited to 8-bit resolution or less.

- The Binary Weighted DAC, which contains one resistor or current source for each bit of the DAC connected to a summing point. These precise voltages or currents sum to the correct output value. This is one of the fastest conversion methods but suffers from poor accuracy because of the high precision required for each individual voltage or current. Such high-precision resistors and current sources are expensive, so this type of converter is usually limited to 8-bit resolution or less.

- The Segmented DAC, which contains an equal resistor or current source segment for each possible value of DAC output. An 8-bit binary-segmented DAC would have 255 segments, and a 16-bit binary-segmented DAC would have 65,535 segments. This is perhaps the fastest and highest precision DAC architecture but at the expense of high cost. Conversion speeds of >1 billion samples per second have been reached with this type of DAC.

- Hybrid DACs, which use a combination of the above techniques in a single converter. Most DAC integrated circuits are of this type due to the difficulty of getting low cost, high speed and high precision in one device.

- This paper will present the 12 bit binary weighted DAC.

- The DAC operates at a maximum frequency of 25MHz using .6um technology to drive a load of 20pF.

Blocks of the DACs

The DAC consists of the following blocks:

1. Binary weighted ladder

2. Op-amp.

Binary Weighted Ladder

The binary weighted ladder consists of 30K and 15K resistors placed in a configuration as shown in figure. The inputs to the ladder are fed from a 12 bit ROM. The input voltages range from 0V to 5V.

The output of the ladder in fed to an OP-AMP. It only supplies a voltage from 3.1V to 4.1 V as an output. Biasing for such an output range was decided by the characteristics of the OP-AMP.

Biasing was implemented using a buffer placed with the ladder network as shown in figure to maintain linearity in the DAC output.

So when all inputs are low, the ladder supplied a voltage of 3.1 V to the OP-AMP and when all the inputs are high a voltage of 4.1V was supplied.

For a 10 bit ladder and 0-5V rails.

The accuracy needed is 5/1025 = 0.0049V,

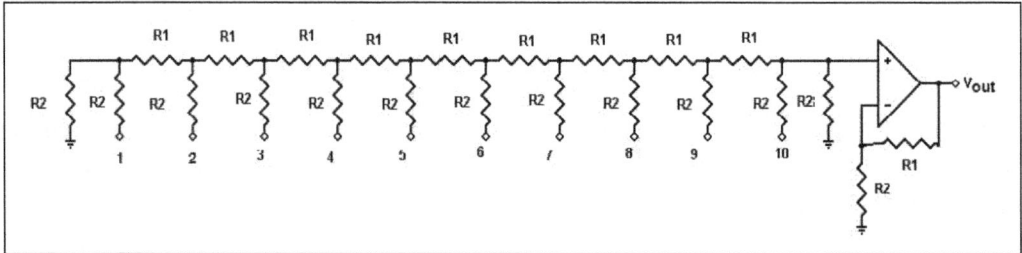

But this is the max voltage variation in both direction of the ideal voltage. Hence the variation from ideal output is 0.0049/2 = 0.00245V.

Which means that for every bit from the input, an analog voltage of 0.0049 is produced at the output.

From the ladder schematic the most stringent accuracy is needed when the input varies from 0111111111 to 1000000000.

For such a transition $V_{out} = V_{ref} R1/R2 = 2.497V$:

R1/R2 = 0.1%

Which is the matching needed in this technology for linearity to be maintained in the output.

Below is shown the schematic of the ladder in two stages:

    1) One with the binary weighted resistors

    2) One with the buffer placed for biasing the op-amp.

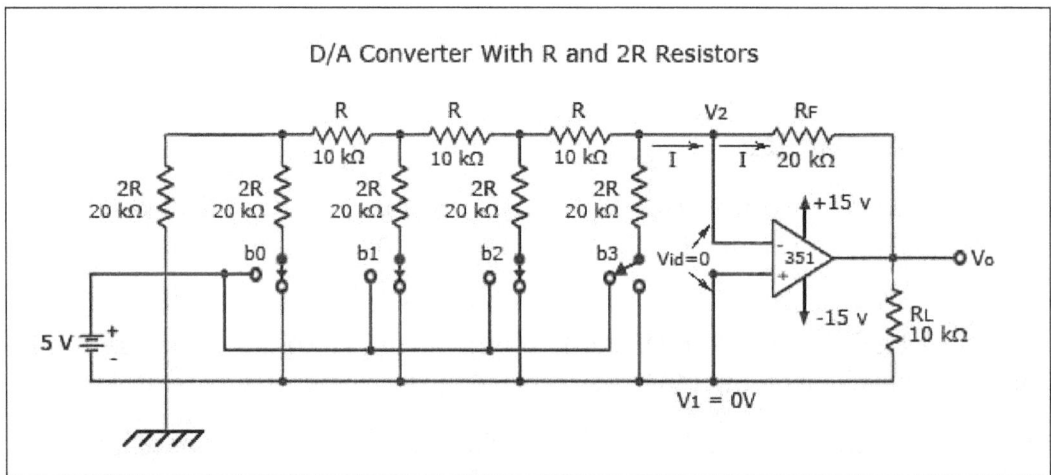

Figure of buffer for bias to op-amp.

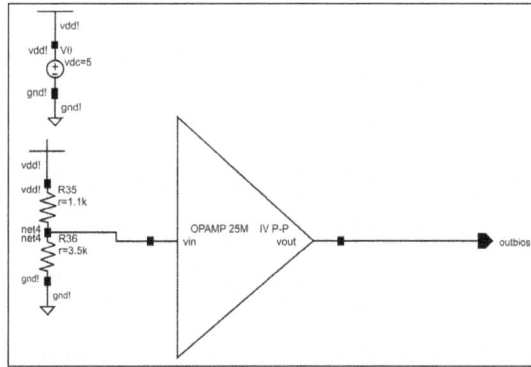

The complete schematic of the ladder including the biasing for the op-amp.

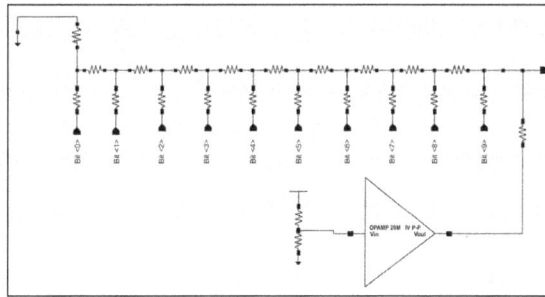

OP-AMP

The design of the op-amp was based on the following design parameters:

- 10 input bits,

- 0-5V rail voltages,

- 50Mhz clock frequency or 25Mhz frequency for DAC.

Open Loop Gain

- The op-amp in this DAC is used in unity feedback configuration.

- The closed loop gain for an op-amp is given by A/(A+1), where A is the open loop gain.

- The error V_{out}/V_{in} is highest at the highest possible V which is V_{ref}=5V.

- Hence A was calculated to be a minimum 2082.

Gain Bandwith

For a maximum operating frequency of 25Mhz, the output needs to be settled in 40ns Again as in the calculation of the open-loop gain, the maximum speed is needed when the output voltage is the highest i.e. 5V.

4.9976 = 5(1-e^(-t/RC))

RC=5.2ns

Or the gain bandwidth comes out to be 20Mhz.

Circuit Design

The op-amp was designed to work in 2 stages:

1) A differential first stage which works as an input for the binary ladder.

2) Final gain stage for an output load of 20pF.

Biasing Circuit

The biasing circuit for the op-amp was implemented using p-mos current mirrors and n-mos current sources for both the stages.

To avoid the affect of slew rate on the op-amp the current to be delivered in the first stage was calculated to be 600uA. The current in the 2nd stage was calculated from the 1st stage to be about 2.3mA. Schematics of the biasing stages are shown in figure.

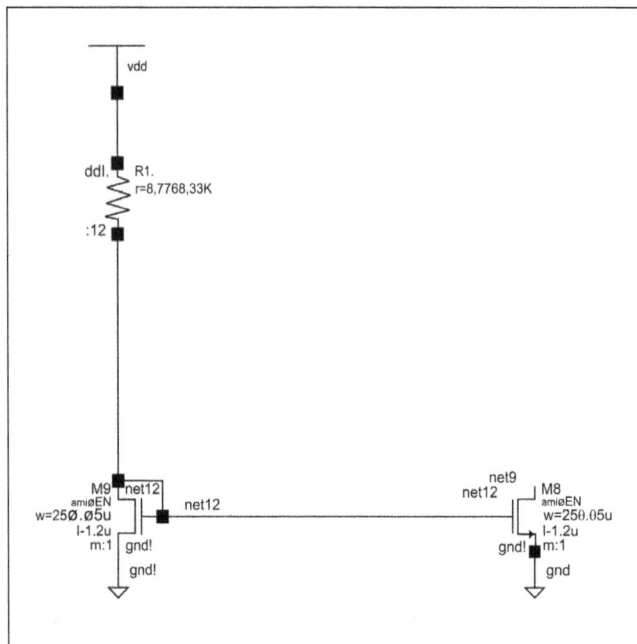

Bias circuit schematic of 1st stage of op-amp

First Stage

The first stage of the op-amp as shown in the figure gives a gain of about 45-60 from a common mode range of 3-4V.

Sizes of the transistors are given in figure.

The lengths of transistors had to be increased to provide high small signal output impedance for a high gain.

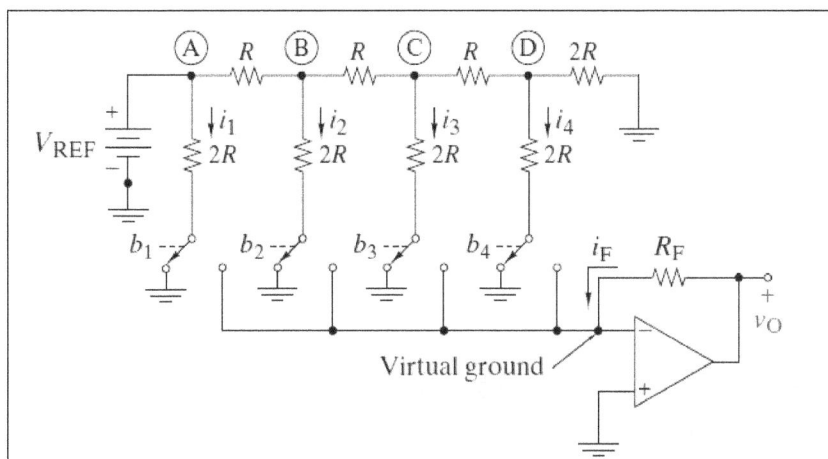

Second Stage

The second stage of the op-amp is acts as a gain stage and to provide enough current to drive a load of 20pF. The compensation capacitor was placed to improve the bandwidth of the amplifier.

Taking parasitic and load capacitance in to account, the maximum frequency of the second pole was calculated to be:

$$1/RC = gm2/(C1+C2)$$

Where, C1 and C2 are given from the figure as below:

$$C1 = Cgd1 + Cgd3 + Cdb1 + Cdb3 + Cgs5$$

$$C2 = Cc + Cdb5$$

$$Cl = Cc + Cgd5$$

$$gm2 = gm5$$

c1+ c2 can be taken as 20pf as parasitics are negligible as compared to load capacitor.

Hence gm2 comes out to be 6.2 mA/V.

And from the formula:

$$gm = 2I/(Vgs-Vt)$$

Where, I= bias current in second stage is calculated to be 2.3 mA.

The 1st pole is arbitrarily chosen to lie at 16Khz.

And the 2nd pole is calculated to be 50Mhz.

A resistor is placed in series with the capacitor so that the zero does not interfere in the transfer function of the gain.

The compensation capacitor is calculated to be 6.44 pF.

And the zero resistor is calculated from the formula:

Rz = (Cl+Cc)/gm5*Cc

Schematic of the 2nd stage:

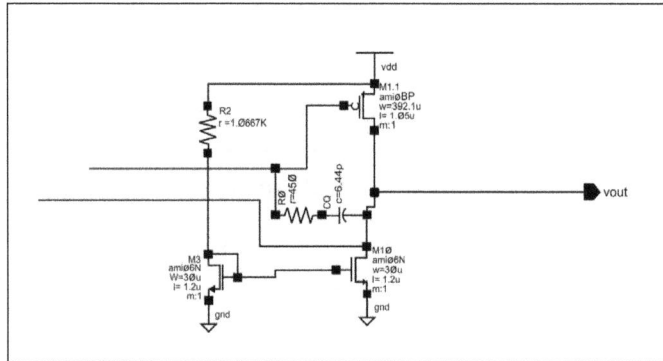

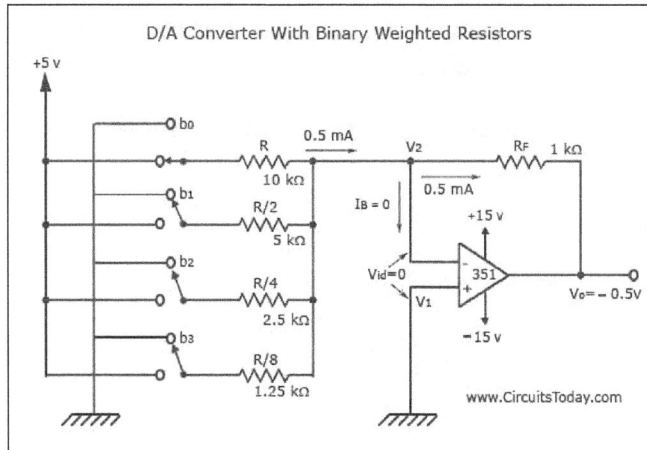

DAC

The complete blocks for the DAC is shown in figure below with test circuit to see the DAC works with linearly increasing signal before testing it with the ROM output.

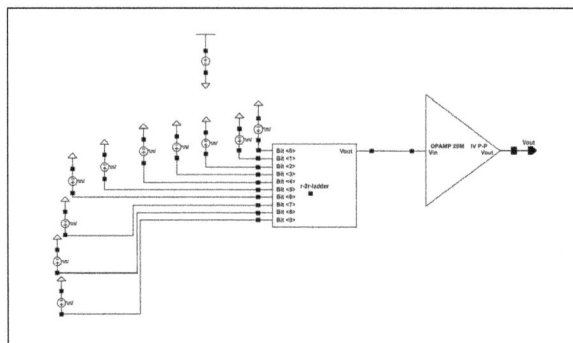

Layout Design

The layout for the DAC is shown in the images below in .6um AMI technology. Guard rings of grounded metal 1 to substrate contacts are used to isolate logic components. Supply decoupling capacitors are built in any space not occupied by circuitry.

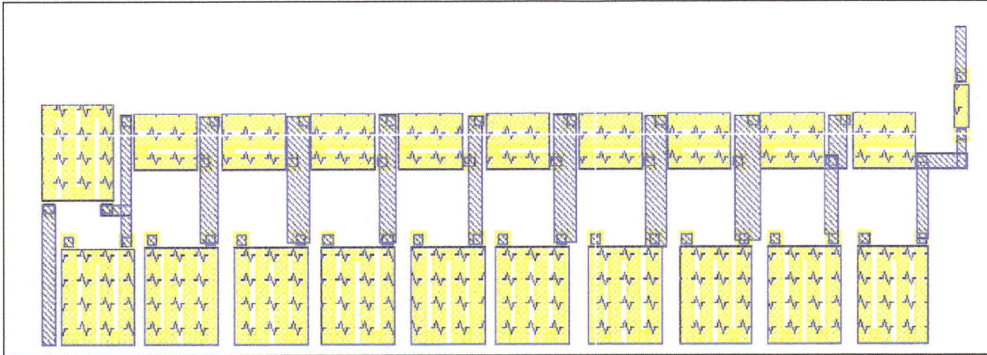

Layout of resistors was done with elec or poly layer

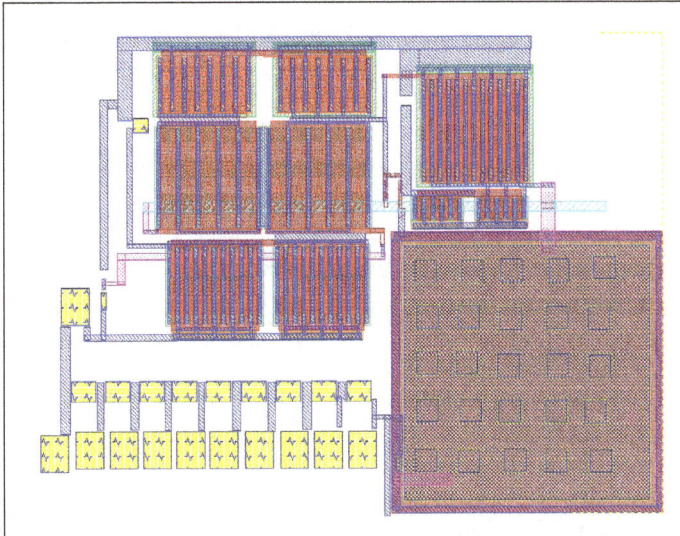

Layout of ladder with buffer shown

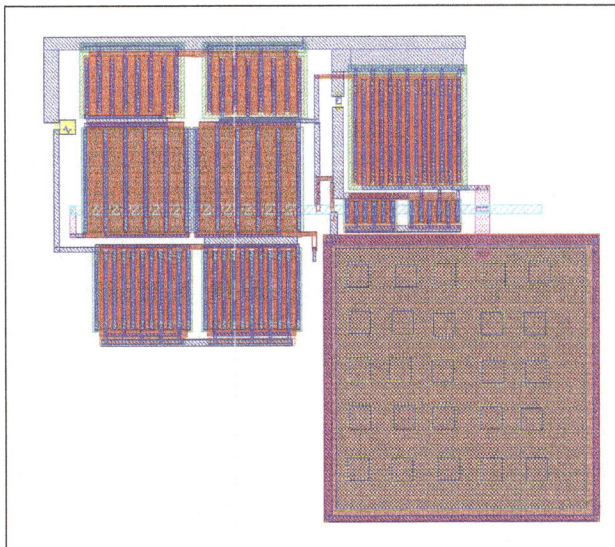

Layout of op-amp with compensating capacitor

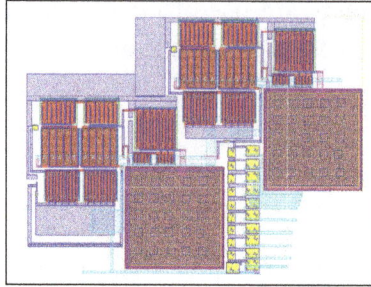

Complete layout of DAC, with 10 inputs fed from ROM

Inverter

An inverter converts the DC voltage to an AC voltage. In most cases, the input DC voltage is usually lower while the output AC is equal to the grid supply voltage of either 120 volts, or 240 Volts depending on the country.

The inverter may be built as standalone equipment for applications such as solar power, or to work as a backup power supply from batteries which are charged separately.

The other configuration is when it is a part of a bigger circuit such as a power supply unit, or a UPS. In this case, the inverter input DC is from the rectified mains AC in the PSU, while from either the rectified AC in the in the UPS when there is power, and from the batteries whenever there is a power failure.

There are different types of inverters based on the shape of the switching waveform. These have varying circuit configurations, efficiencies, advantages and disadvantages

An inverter provides an ac voltage from dc power sources and is useful in powering electronics and electrical equipment rated at the ac mains voltage. In addition they are widely used in the switched mode power supplies inverting stages. The circuits are classified according the switching technology and switch type, the waveform, the frequency and output waveform.

Basic Inverter Operation

The basic circuits include an oscillator, control circuit, drive circuit for the power devices, switching devices, and a transformer.

The conversion of dc to alternating voltage is achieved by converting energy stored in the dc source such as the battery, or from a rectifier output, into an alternating voltage. This is done using switching devices which are continuously turned on and off, and then stepping up using the transformer. Although there are some configurations which do not use a transformer, these are not widely used.

The DC input voltage is switched on and off by the power devices such as MOSFETs or power transistors and the pulses fed to the primary side of the transformer. The varying voltage in the primary induces an alternating voltage at secondary winding. The transformer also works as an amplifier where it increases the output voltage at a ratio determined by the turn's ratio. In most cases the output voltage is raised from the standard 12 volts supplied by the batteries to either 120 Volts or 240 volts AC.

The three commonly used Inverter output stages are, a push-pull with centre tap transformer, push-pull half-bridge, or push-pull full bridge. The push pull with centre tap is most popular due to its simplicity and, guaranteed results; however, it uses a heavier transformer and has a lower efficiency.

A simple push pull DC to AC inverter with centre tap transformer circuit is a shown in the figure.

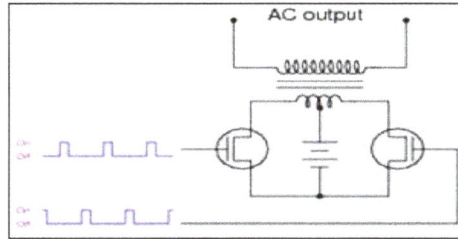

Basic inverter switching circuit

Inverter Output Waveforms

The inverters are classified according to their output waveforms with the three common types being the square wave, the pure sine wave and the modified sine wave.

The square wave is simple and cheaper, however, it has a low power quality compared to the other two. The modified square wave provides a better power quality (THD~ 45%) and is suitable for most electronic equipment. These have rectangular pulses that have dead spots between the positive half cycle and the negative half cycle (THD about 24%).

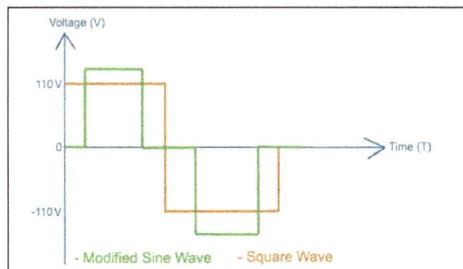

Modified sine waveform

The true sine wave inverter has the best waveform with the lowest THD of about 3%. However, it is the most expensive and used in applications such as medical equipment, stereos, laser printers and other applications requiring sinusoidal waveforms. These are also used in the grid ties inverters and grid connected equipment.

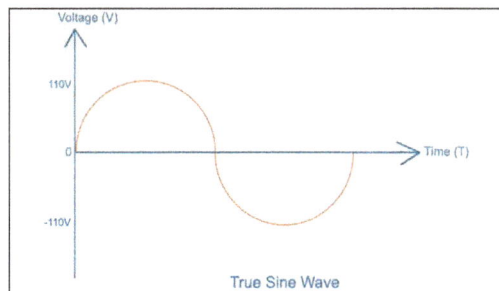

Pure sine wave

Applications

Inverters are used for a variety of applications that range from small car adapters to household or office applications, and large grid systems:

- Uninterruptible power supplies.

- As standalone inverters.

- In solar power systems.

- As a building block of a switched mode power supply.

Rectifier

A rectifier is an electrical device that converts an Alternating Current (AC) into a Direct Current (DC) by using one or more P-N junction diodes.

When the voltage is applied to the P-N junction diode in such a way that the positive terminal of the battery is connected to the p-type semiconductor and the negative terminal of the battery is connected to the n-type semiconductor, the diode is said to be forward biased.

When this forward bias voltage is applied to the P-N junction diode, a large number of free electrons (majority carriers) in the n-type semiconductor experience a repulsive force from the negative terminal of the battery similarly a large number of holes (majority carriers) in the p-type semiconductor experience a repulsive force from the positive terminal of the battery.

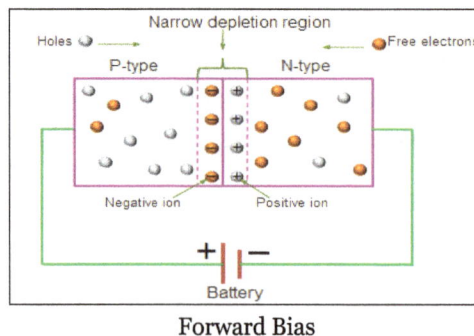

Forward Bias

As a result, the free electrons in the n-type semiconductor start moving from n-side to p-side similarly the holes in the p-type semiconductor start moving from p-side to n-side.

We know that electric current means the flow of charge carriers (free electrons and holes). Therefore, the flow of electrons from n-side to p-side and the flow of holes from p-side to n-side conduct electric current. The majority carriers produce the electric current in forward bias condition. So the electric current produced in forward bias condition is also known as majority current.

When the voltage is applied to the P-N junction diode in such a way that the positive terminal of the battery is connected to the n-type semiconductor and the negative terminal of the battery is connected to the p-type semiconductor, the diode is said to be reverse biased.

When this reverse bias voltage is applied to the P-N junction diode, a large number of free electrons (majority carriers) in the n-type semiconductor experience an attractive force from the positive terminal of the battery similarly a large number of holes (majority carriers) in the p-type semiconductor experience an attractive force from the negative terminal of the battery.

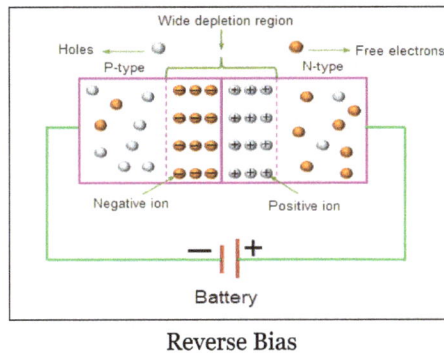

Reverse Bias

As a result, the free electrons (majority carriers) in the n-type semiconductor moves away from the P-N junction and attracted to the positive terminal of the battery similarly the holes (majority carriers) in the p-type semiconductor moves away from the P-N junction and attracted to the negative terminal of the battery.

Therefore, the electric current flow does not occur across the P-N junction. However, the minority carriers (free electrons) in the p-type semiconductor experience a repulsive force from the negative terminal of the battery similarly the minority carriers (holes) in the n-type semiconductor experience a repulsive force from the positive terminal of the battery.

As a result, the minority carriers free electrons in the p-type semiconductor and the minority carriers holes in the n-type semiconductor starts flowing across the junction. Thus, electric current is produced in the reverse bias diode due to the minority carriers. However, the electric current produced by the minority carriers is very small. So the minority carrier current in the reverse bias condition is neglected.

Thus, the P-N junction diode allows electric current in forward bias condition and blocks electric current in reverse bias condition. In simple words, a P-N junction diode allows electric current in only one direction. This unique property of the diode allows it to acts like a rectifier.

The forward bias and reverse bias voltage applied to the diode is nothing but a DC voltage. A DC voltage produces a current which always flows in one direction (either forward direction or backward direction).

But an AC voltage produces a current which always reverses its direction many times a second (forward to backward and backward to forward).

We have observed how a diode behaves when DC voltage (forward bias voltage and reverse bias voltage) is applied to it. Now let's take look at a P-N junction diode when AC voltage is applied to it.

The AC voltage or AC current is often represented by a sinusoidal waveform whereas the DC current is represented by a straight horizontal line.

In the sinusoidal waveform, the upper half cycle represents the positive half cycle and the lower half cycle represents the negative half cycle.

The positive half cycle of the AC voltage is analogous to the forward bias DC voltage and the negative half cycle of the AC voltage is analogous to the reverse bias DC voltage.

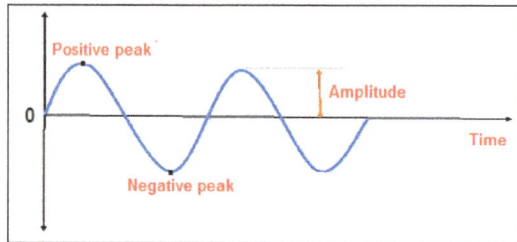

The alternating current starts from zero and grows to peak forward current or peak positive current. The positive peak of the sinusoidal waveform represents the maximum or peak forward current. After reaching the peak forward current, it starts decreasing and reaches to zero.

After a short period, the alternating current starts increasing in the reverse or negative direction and grows to peak reverse current or peak negative current. The negative peak of the sinusoidal waveform represents the maximum or peak reverse current. After reaching the peak reverse current, it starts decreasing and reaches to zero. Likewise, the alternating current continuously changes its direction in a short period.

When AC voltage or AC current is applied across the P-N junction diode, during the positive half cycle the diode is forward biased and allows electric current through it. However, when the AC current reverses its direction to negative half cycle, the diode is reverse biased and does not allow electric current through it. In simple words, during the positive half cycle, the diode allows current and during the negative half cycle, the diode blocks current. Thus, electric current flows through the diode only during the positive half cycle of the AC current.

This current which flows across the diode is nothing but a DC current. Thus, the P-N junction diode acts like a rectifier by converting the AC current into DC current.

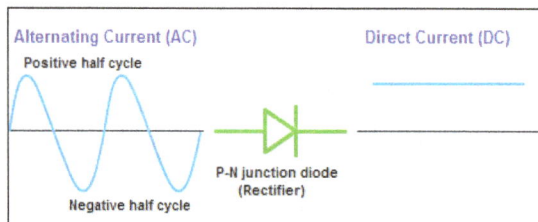

However, the DC current produced by a basic rectifier (half wave rectifier) is not a pure DC current. It is a pulsating DC current.

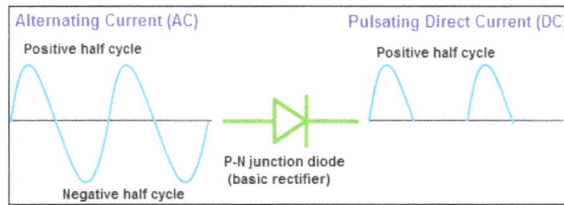

The pulsating direct current is a type of DC current whose value changes over a short period.

The pulsating DC current starts from zero and grows to the maximum forward current (peak level), and decreases to zero. However, the pulsating DC current does not change its direction periodically like AC current.

The pulsating DC current always flows in one direction like the pure DC current. However, the value of pulsating DC current or pulsating DC voltage slightly changes over a given period. The electric current produced by batteries, power supplies, and solar panels is a pure DC current.

By using the combination of components such as capacitors, inductors, and resistors in the circuit, we can achieve the smoothening of pulsating DC to pure DC.

Types of Rectifiers

Half Wave Rectifier

A half wave rectifier is defined as a type of rectifier that only allows one half-cycle of an AC voltage waveform to pass, blocking the other half-cycle. Half-wave rectifiers are used to convert AC voltage to DC voltage, and only require a single diode to construct.

A rectifier is a device that converts alternating current (AC) to direct current (DC). It is done by using a diode or a group of diodes. Half wave rectifiers use one diode, while a full wave rectifier uses multiple diodes.

The working of a half wave rectifier takes advantage of the fact that diodes only allow current to flow in one direction.

Half Wave Rectifier Theory

A half wave rectifier is the simplest form of rectifier available. The diagram below illustrates the basic principle of a half-wave rectifier. When a standard AC waveform is passed through a half-wave rectifier, only half of the AC waveform remains. Half-wave rectifiers only allow one half-cycle (positive or negative half-cycle) of the AC voltage through and will block the other half-cycle on the DC side, as seen below.

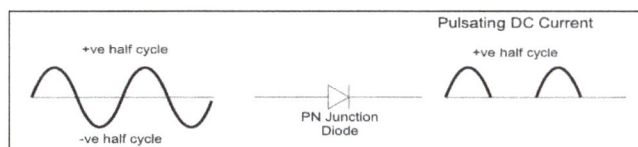

Only one diode is required to construct a half-wave rectifier. In essence, this is all that the half-wave rectifier is doing.

Since DC systems are designed to have current flowing in a single direction (and constant voltage – which we'll describe later), putting an AC waveform with positive and negative cycles through a DC device can have destructive (and dangerous) consequences. So we use half-wave rectifiers to convert the AC input power into DC output power.

But the diode is only part of it – a complete half-wave rectifier circuit consists of 3 main parts:

1. A transformer,

2. A resistive load,

3. A diode.

A half wave rectifier circuit diagram looks like this:

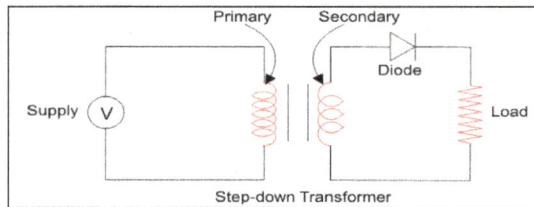

We'll now go through the process of how a half-wave rectifier converts an AC voltage to a DC output.

First, a high AC voltage is applied to the to the primary side of the step-down transformer and we will get a low voltage at the secondary winding which will be applied to the diode.

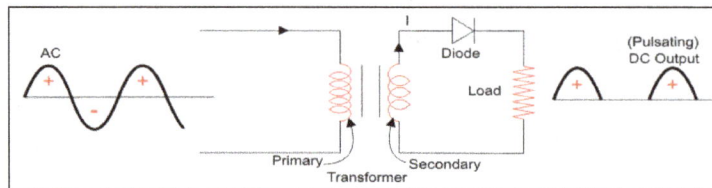

During the positive half cycle of the AC voltage, the diode will be forward biased and the current flows through the diode. During the negative half cycle of the AC voltage, the diode will be reverse biased and the flow of current will be blocked. The final output voltage waveform on the secondary side (DC) is shown in figure above.

This can be confusing on first glance – so let's dig into the theory of this a bit more.

We'll focus on the secondary side of the circuit. If we replace the secondary transformer coils with a source voltage, we can simplify the circuit diagram of the half-wave rectifier as:

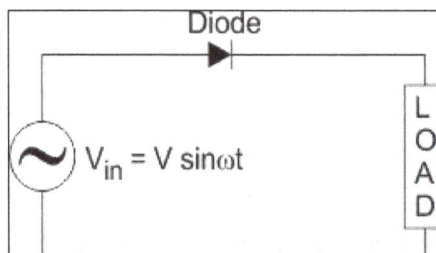

Now we don't have the transformer part of the circuit distracting us.

For the positive half cycle of the AC source voltage, the equivalent circuit effectively becomes:

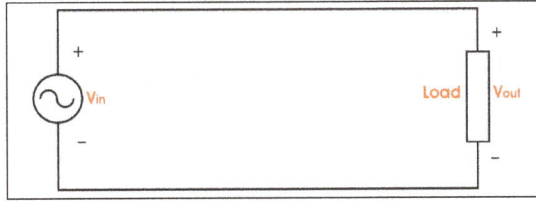

This is because the diode is forward biased, and is hence allowing current to pass through. So we have a closed circuit.

But for the negative half cycle of the AC source voltage, the equivalent circuit becomes:

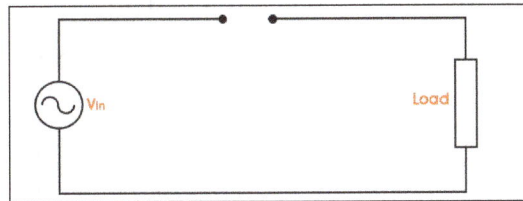

Because the diode is now in reverse bias mode, no current is able to pass through it. As such, we now have an open circuit. Since current cannot flow through to the load during this time, the output voltage is equal to zero.

This all happens very quickly – since an AC waveform will oscillate between positive and negative many times each second (depending on the frequency).

Here's what the half wave rectifier waveform looks like on the input side (V_{in}), and what it looks like on the output side (V_{out}) after rectification (i.e. conversion from AC to DC):

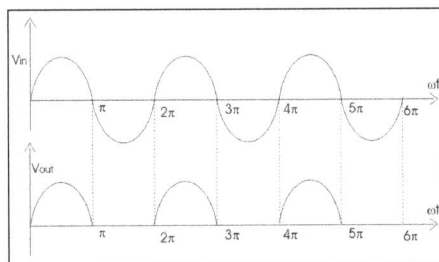

The graph actually shows a positive half wave rectifier. This is a half-wave rectifier which only allows the positive half-cycles through the diode, and blocks the negative half-cycle.

The voltage waveform before and after a positive half wave rectifier is shown in figure.

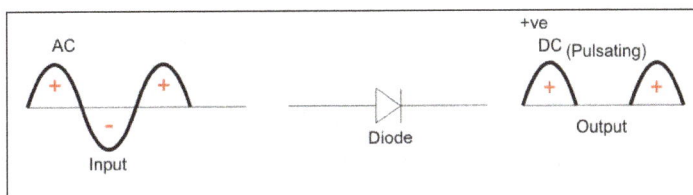

Conversely, a negative half-wave rectifier will only allow negative half-cycles through the diode and will block the positive half-cycle. The only difference between a posiveand negative half wave rectifier is the direction of the diode.

As you can see in figure below, the diode is now in the opposite direction. Hence the diode will now be forward biased only when the AC waveform is in its negative half cycle.

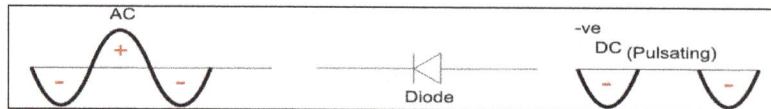

Half Wave Rectifier Capacitor Filter

The output waveform we have obtained from the theory above is a pulsating DC waveform. This is what is obtained when using a half wave rectifier without a filter.

Filters are components used to convert (smoothen) pulsating DC waveforms into constant DC waveforms. They achieve this by suppressing the DC ripples in the waveform.

Although half-wave rectifiers without filters are theoretically possible, they can't be used for any practical applications. As DC equipment requires a constant waveform, we need to 'smooth out' this pulsating waveform for it to be any use in the real world.

This is why in reality we use half wave rectifiers with a filter. A capacitor or an inductor can be used as a filter – but half wave rectifier with capacitor filter is most commonly used.

The circuit diagram below shows how a capacitive filter is can be used to smoothen out a pulsating DC waveform into a constant DC waveform.

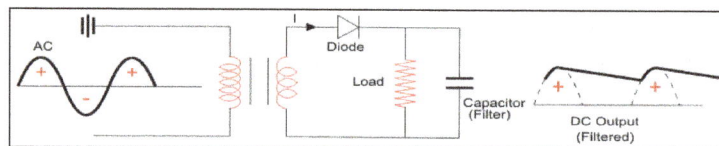

Half Wave Rectifier Formula

We will now derive the various formulas for a half wave rectifier based on the preceding theory and graphs.

Ripple Factor of Half Wave Rectifier

'Ripple' is the unwanted AC component remaining when converting the AC voltage waveform into a DC waveform. Even though we try out best to remove all AC components, there is still some small amount left on the output side which pulsates the DC waveform. This undesirable AC component is called 'ripple'.

To quantify how well the half-wave rectifier can convert the AC voltage into DC voltage, we use what is known as the ripple factor (represented by γ or r). The ripple factor is the ratio between the RMS value of the AC voltage (on the input side) and the DC voltage (on the output side) of the rectifier.

The formula for ripple factor is:

$$\gamma = \sqrt{\left(\frac{V_{rms}}{V_{DC}}\right)^2 - 1}$$

Which can also be rearranged to equal:

$$Ripple\ factor\ (r) = \frac{\left(I_{rms}^2 - I_{dc}^2\right)}{I_{dc}} = 1.21$$

The ripple factor of half wave rectifier is equal to 1.21 (i.e. $\gamma = 1.21$).

Note that for us to construct a good rectifier, we want to keep the ripple factor as low as possible. This is why we use capacitors and inductors as filters to reduce the ripples in the circuit.

Efficiency of Half Wave Rectifier

Rectifier efficiency (η) is the ratio between the output DC power and the input AC power. The formula for the efficieny is equal to:

$$\eta = \frac{P_{dc}}{P_{ac}}$$

The efficiency of a half wave rectifier is equal to 40.6% (i.e. $\eta_{max} = 40.6\%$).

RMS value of Half Wave Rectifier

To derive the RMS value of half wave rectifier, we need to calculate the current across the load. If the instantaneous load current is equal to $i_L = I_m \sin\omega t$, then the average of load current (I_{DC}) is equal to:

$$I_{dc} = \frac{1}{2\pi} \int_0^{\pi} I_m \sin \omega t = \frac{I_m}{\pi}$$

Where I_m is equal to the peak instantaneous current across the load (I_{max}). Hence the output DC current (I_{DC}) obtained across the load is:

$$I_{dc} = \frac{I_{max}}{\pi},$$

Where,

I_{max}=maximum amplitude of DC current.

For a half-wave rectifier, the RMS load current (I_{rms}) is equal to the average current (I_{DC}) multiple by $\pi/2$. Hence the RMS value of the load current (I_{rms}) for a half wave rectifier is:

$$I_{rms} = \frac{I_m}{4}$$

Where,

$I_m = I_{max}$ which is equal to the peak instantaneous current across the load.

Peak Inverse Voltage of Half Wave Rectifier

Peak Inverse Voltage (PIV) is the maximum voltage that the diode can withstand during reverse bias condition. If a voltage is applied more than the PIV, the diode will be destroyed.

Form Factor of Half Wave Rectifier

Form factor (F.F) is the ratio between RMS value and average value, as shown in the formula below:

$$F.F = \frac{RMS\ value}{Average\ value}$$

The form factor of a half wave rectifier is equal to 1.57 (i.e. F.F= 1.57).

Output DC Voltage

The output voltage (V_{DC}) across the load resistor is denoted by:

$$V_{DC} = \frac{V_{S\,max}}{\pi},$$

Where,

Vs_{max}= maximum amplitude of secondary voltage.

Applications of Half Wave Rectifier

Half wave rectifiers are not as commonly used as full-wave rectifiers. Despite this, they still have some uses:

- For rectification applications.

- For signal demodulation applications.

- For signal peak applications.

Advantages of Half Wave Rectifier

The main advantage of half-wave rectifiers is in their simplicity. As they don't require as many components, they are simpler and cheaper to setup and construct. As such, the main advantages of half-wave rectifiers are:

- Simple (lower number of components).

- Cheaper up front cost (as their is less equipment. Although there is a higher cost over time due to increased power losses).

Disadvantages of Half Wave Rectifier

The disadvantages of half-wave rectifiers are:

- They only allow a half-cycle through per sinewave, and the other half-cycle is wasted. This leads to power loss.

- They produces a low output voltage.

- The output current we obtain is not purely DC, and it still contains a lot of ripple (i.e. it has a high ripple factor).

3 Phase Half Wave Rectifier

All of the theory above has dealt with a single phase half wave rectifier. Although the principle of a 3 phase half wave rectifier is the same, the characteristics are different. The waveform, ripple factor, efficiency, and RMS output values are not the same.

The three phase half wave rectifier is used for conversion of three-phase AC power to DC power. Here the switches are diodes, and hence they are uncontrolled switches. That is to say, there is no way of controlling the on and off times of these switches.

The 3 phase half wave diode rectifier is generally constructed with a three-phase supply connected to a three-phase transformer where the secondary winding of the transformer is always connected via star connection. This is because the neutral point is required to connect the load back to the transformer secondary windings, providing a return path for the flow of power.

A typical configuration of a three-phase half wave rectifier supplying to a purely resistive load is shown below. Here, each phase of the transformer is considered as an individual alternating source. The simulation and measurement of voltages are as shown in the circuit below. Here we have connected an individual voltmeter across each source as well as across the load.

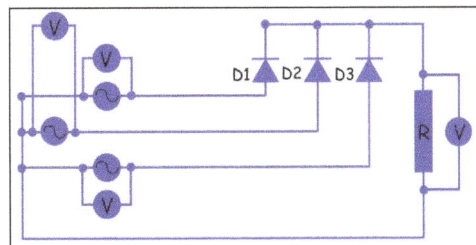

The three-phase voltages are shown below.

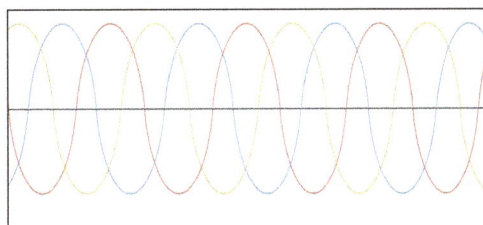

The voltage across the resistive load is shown below. The voltage is shown in black.

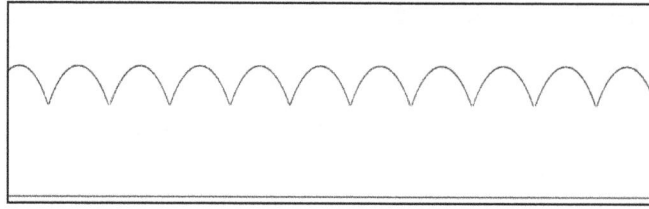

So we can see from the above figure that the diode D1 conducts when the R phase has a value of the voltage that is higher than the value of the voltage of the other two phases, and this condition begins when the R phase is at a 30° and repeats after every complete cycle. That is to say, the next time diode DI begins to conduct is at 390°. Diode D2 takes over conduction from D1 which stops conducting at angle 150° because at this instant the value of voltage in B phase becomes higher than the voltages in the other two phases. So each diode conducts for an angle of 150° − 30° = 120°.

Here, the waveform of the resulting DC voltage signal is not purely DC as it is not flat, but rather it contains a ripple. And the frequency of the ripple is 3 × 50 = 150 Hz.

The average of the output voltage across the resistive load is given by:

$$V_o = \frac{1}{\frac{2\pi}{3}} \int_{\frac{\pi}{6}}^{\frac{5\pi}{6}} V_{m_{phase}} \sin(\omega t) d(\omega t)$$

$$= \frac{3\sqrt{6}}{2\pi} V_{phase} = 1.17 \, V_{phase} = 0.827 V_{m_{phase}} = \frac{3}{2} V_{m_{line}}$$

Where,

$$V_{m_{phase}} = \sqrt{2} V_{phase}$$

$$V_{m_{line}} = \sqrt{3} V_{m_{Phase}} = \sqrt{6} V_{phase}$$

The RMS value of the output voltage is given by:

$$V_{o_{rms}} = \left[\frac{1}{\frac{2\pi}{3}} \int_{\frac{\pi}{6}}^{\frac{5\pi}{6}} \left(V_{m_{Phase}} \sin(\omega t) \right)^2 d(\omega t) \right]^{\frac{1}{2}}$$

$$= 0.84068 \, V_{m_{phase}}$$

The ripple voltage is equal to:

$$V_r = \sqrt{V_{o_{rms}}^2 - V_o^2}$$

$$= \sqrt{0.84068^2 - 0.827^2} \, V_{m_{phase}}$$

$$= 0.151 V_{m_{phase}}$$

And the voltage ripple factor is equal to: $\dfrac{V_r}{V_o} = \dfrac{0.151}{0.827} = 0.1826 = 18.26\%$.

The equation above shows that the voltage ripple is significant. This is undesirable as this leads to unnecessary power loss.

DC output power:

$$P_o = V_o I_o = \frac{3\sqrt{3}}{2\pi} V_{m_{phase}} \frac{3\sqrt{3}}{2\pi} I_{m_{phase}}$$

AC input power:

$$P_i = V_{or} I_{or} = (0.84068)^2 V_{m_{phase}} \, I_{m_{phase}}$$
$$= 0.7067 \, V_{m_{phase}} \, I_{m_{phase}}$$

Efficiency:

$$\eta = \frac{P_o}{P_i} = \frac{0.684 \, V_{m_{phase}} \, I_{m_{phase}}}{0.7067 \, V_{m_{phase}} \, I_{m_{phase}}} = 0.968 = 96.8\%$$

Even though the efficiency of the 3 phase half-wave rectifier is seemingly high, it is still less than the efficiency provided by a 3 phase full wave diode rectifier. Although three phase half wave rectifiers are cheaper, this cost saving is insignificant compared to the money lost in their higher power losses. As such, three-phase half-wave rectifiers are not commonly used in industry.

Full Wave Rectifier

A full wave rectifier circuit produces an output voltage or current which is purely DC or has some specified DC component. Full wave rectifiers have some fundamental advantages over their half wave rectifier counterparts. The average (DC) output voltage is higher than for half wave, the output of the full wave rectifier has much less ripple than that of the half wave rectifier producing a smoother output waveform.

In a Full Wave Rectifier circuit two diodes are now used, one for each half of the cycle. A multiple winding transformer is used whose secondary winding is split equally into two halves with a common centre tapped connection, (C). This configuration results in each diode conducting in turn when its anode terminal is positive with respect to the transformer centre point C producing an output during both half-cycles twice that for the half wave rectifier so it is 100% efficient.

Full Wave Rectifier Circuit

The full wave rectifier circuit consists of two *power diodes* connected to a single load resistance (R_L) with each diode taking it in turn to supply current to the load. When point A of the transformer is positive with respect to point C, diode D_1 conducts in the forward direction as indicated by the arrows.

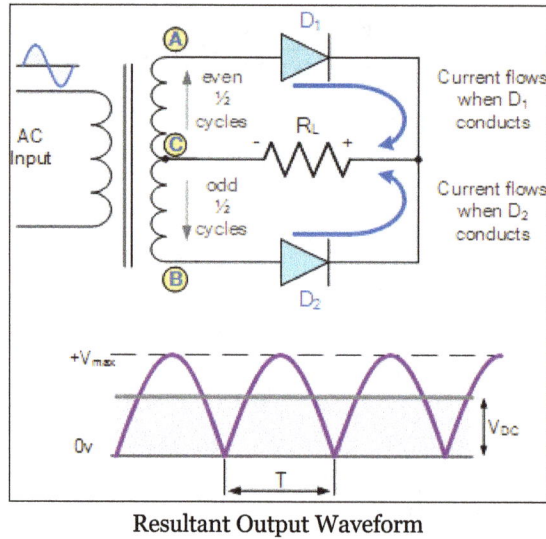

Resultant Output Waveform

When point B is positive (in the negative half of the cycle) with respect to point C, diode D_2 conducts in the forward direction and the current flowing through resistor R is in the same direction for both half-cycles. As the output voltage across the resistor R is the phasor sum of the two waveforms combined, this type of full wave rectifier circuit is also known as a "bi-phase" circuit.

Partsim Simulation Waveform

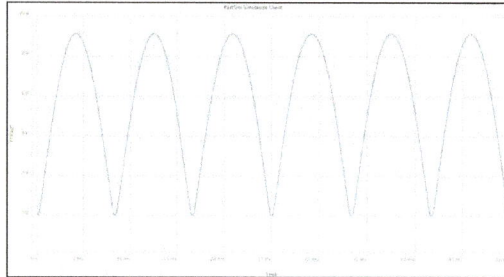

As the spaces between each half-wave developed by each diode is now being filled in by the other diode the average DC output voltage across the load resistor is now double that of the single half-wave rectifier circuit and is about $0.637V_{max}$ of the peak voltage, assuming no losses.

$$V_{d.c.} = \frac{2V_{max}}{\pi} = 0.637\ V_{max} = 0.9\ V_{RMS}$$

Where,

V_{MAX} is the maximum peak value in one half of the secondary winding and V_{RMS} is the rms value.

The peak voltage of the output waveform is the same as before for the half-wave rectifier provided each half of the transformer windings have the same rms voltage value. To obtain a different DC voltage output different transformer ratios can be used.

The main disadvantage of this type of full wave rectifier circuit is that a larger transformer for a given power output is required with two separate but identical secondary windings making this type of full wave rectifying circuit costly compared to the "Full Wave Bridge Rectifier" circuit equivalent.

The Full Wave Bridge Rectifier

Another type of circuit that produces the same output waveform as the full wave rectifier circuit above is that of the Full Wave Bridge Rectifier. This type of single phase rectifier uses four individual rectifying diodes connected in a closed loop "bridge" configuration to produce the desired output.

The main advantage of this bridge circuit is that it does not require a special centre tapped transformer, thereby reducing its size and cost. The single secondary winding is connected to one side of the diode bridge network and the load to the other side.

The Diode Bridge Rectifier

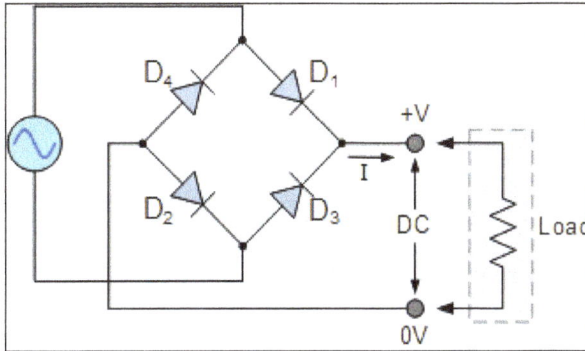

The four diodes labelled D_1 to D_4 are arranged in "series pairs" with only two diodes conducting current during each half cycle. During the positive half cycle of the supply, diodes D_1 and D_2 conduct in series while diodes D_3 and D_4 are reverse biased and the current flows through the load as shown below.

The Positive Half-cycle

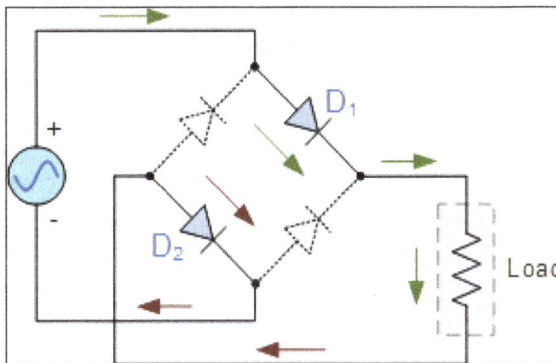

During the negative half cycle of the supply, diodes D_3 and D_4 conduct in series, but diodes D_1 and D_2 switch "OFF" as they are now reverse biased. The current flowing through the load is the same direction as before.

The Negative Half-cycle

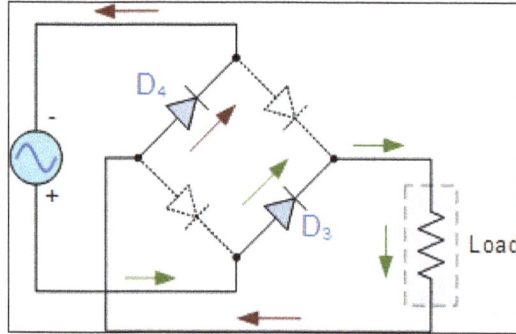

As the current flowing through the load is unidirectional, so the voltage developed across the load is also unidirectional therefore the average DC voltage across the load is $0.637V_{max}$.

Typical Bridge Rectifier

However in reality, during each half cycle the current flows through two diodes instead of just one so the amplitude of the output voltage is two voltage drops ($2*0.7 = 1.4V$) less than the input V_{MAX} amplitude. The ripple frequency is now twice the supply frequency (e.g. 100Hz for a 50Hz supply or 120Hz for a 60Hz supply.)

Although we can use four individual power diodes to make a full wave bridge rectifier, pre-made bridge rectifier components are available "off-the-shelf" in a range of different voltage and current sizes that can be soldered directly into a PCB circuit board or be connected by spade connectors.

The image to the right shows a typical single phase bridge rectifier with one corner cut off. This cut-off corner indicates that the terminal nearest to the corner is the positive or +veoutput terminal or lead with the opposite (diagonal) lead being the negative or -ve output lead. The other two connecting leads are for the input alternating voltage from a transformer secondary winding.

The Smoothing Capacitor

The single phase half-wave rectifier produces an output wave every half cycle and that it was not practical to use this type of circuit to produce a steady DC supply. The full-wave bridge rectifier however, gives us a greater mean DC value (0.637 V_{max}) with less superimposed ripple while the output waveform is twice that of the frequency of the input supply frequency.

We can improve the average DC output of the rectifier while at the same time reducing the AC variation of the rectified output by using smoothing capacitors to filter the output waveform. Smoothing or reservoir capacitors connected in parallel with the load across the output of the full wave bridge rectifier circuit increases the average DC output level even higher as the capacitor acts like a storage device.

Full-wave Rectifier with Smoothing Capacitor

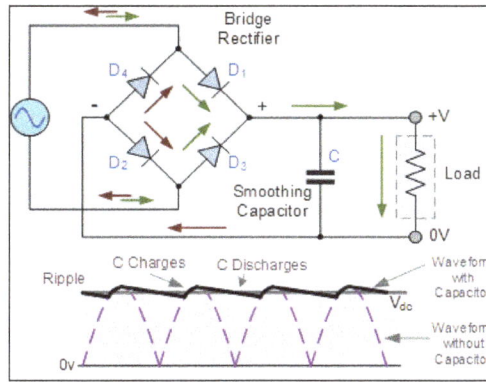

Resultant Waveform Output

The smoothing capacitor converts the full-wave rippled output of the rectifier into a more smooth DC output voltage.

5uF Smoothing Capacitor

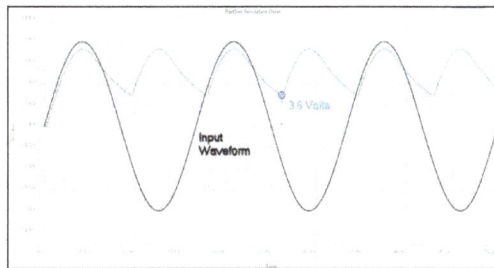

The blue plot on the waveform shows the result of using a 5.0uF smoothing capacitor across the rectifiers output. Previously the load voltage followed the rectified output waveform down to zero volts. Here the 5uF capacitor is charged to the peak voltage of the output DC pulse, but when it drops from its peak voltage back down to zero volts, the capacitor cannot discharge as quickly due to the RC time constant of the circuit.

This result in the capacitor discharging down to about 3.6 volts, in this example, maintaining the voltage across the load resistor until the capacitor re-charges once again on the next positive slope of the DC pulse. In other words, the capacitor only has time to discharge briefly before the next DC pulse recharges it back up to the peak value. Thus, the DC voltage applied to the load resistor drops only by a small amount. But we can improve this still by increasing the value of the smoothing capacitor as shown.

50uF Smoothing Capacitor

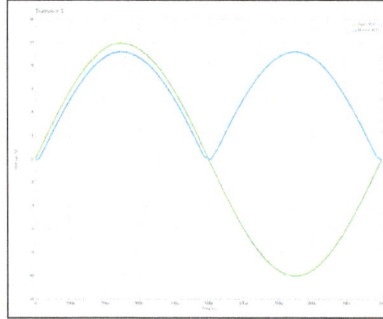

Here we have increased the value of the smoothing capacitor ten-fold from 5uF to 50uF which has reduced the ripple increasing the minimum discharge voltage from the previous 3.6 volts to 7.9 volts.

The effect of a supplying a heavy load with a single smoothing or reservoir capacitor can be reduced by the use of a larger capacitor which stores more energy and discharges less between charging pulses. Generally for DC power supply circuits the smoothing capacitor is an Aluminium Electrolytic type that has a capacitance value of 100uF or more with repeated DC voltage pulses from the rectifier charging up the capacitor to peak voltage.

However, there are two important parameters to consider when choosing a suitable smoothing capacitor and these are its *Working Voltage*, which must be higher than the no-load output value of the rectifier and its *Capacitance Value*, which determines the amount of ripple that will appear superimposed on top of the DC voltage.

Too low a capacitance value and the capacitor has little effect on the output waveform. But if the smoothing capacitor is sufficiently large enough (parallel capacitors can be used) and the load current is not too large, the output voltage will be almost as smooth as pure DC. As a general rule of thumb, we are looking to have a ripple voltage of less than 100mV peak to peak.

The maximum ripple voltage present for a Full Wave Rectifier circuit is not only determined by the value of the smoothing capacitor but by the frequency and load current, and is calculated as:

Bridge Rectifier Ripple Voltage:

$$V_{(ripple)} = \frac{I_{(load)}}{f \times C}, Volts$$

Where,

I is the DC load current in amps, f is the frequency of the ripple or twice the input frequency in Hertz, and C is the capacitance in Farads.

The main advantages of a full-wave bridge rectifier is that it has a smaller AC ripple value for a given load and a smaller reservoir or smoothing capacitor than an equivalent half-wave rectifier. Therefore, the fundamental frequency of the ripple voltage is twice that of the AC supply frequency (100Hz) where for the half-wave rectifier it is exactly equal to the supply frequency (50Hz).

The amount of ripple voltage that is superimposed on top of the DC supply voltage by the diodes can be virtually eliminated by adding a much improved π-filter (pi-filter) to the output terminals of the bridge rectifier. This type of low-pass filter consists of two smoothing capacitors, usually of the same value and a choke or inductance across them to introduce a high impedance path to the alternating ripple component.

Another more practical and cheaper alternative is to use an off the shelf 3-terminal voltage regulator IC, such as a LM78xx (where "xx" stands for the output voltage rating) for a positive output voltage or its inverse equivalent the LM79xx for a negative output voltage which can reduce the ripple by more than 70dB (Datasheet) while delivering a constant output current of over 1 amp.

Bridge Rectifier

A bridge rectifier is a type of full wave rectifier which uses four or more diodes in a bridge circuit configuration to efficiently convert the Alternating Current (AC) into Direct Current (DC).

Bridge Rectifier Construction

The construction diagram of a bridge rectifier is shown in the below figure. The bridge rectifier is made up of four diodes namely D_1, D_2, D_3, D_4 and load resistor R_L. The four diodes are connected in a closed loop (Bridge) configuration to efficiently convert the Alternating Current (AC) into Direct Current (DC). The main advantage of this bridge circuit configuration is that we do not require an expensive center tapped transformer, thereby reducing its cost and size.

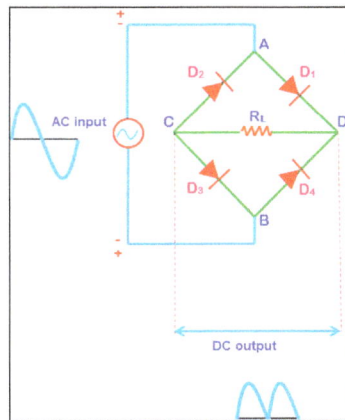

Bridge Rectifier

The input AC signal is applied across two terminals A and B and the output DC signal is obtained across the load resistor R_L which is connected between the terminals C and D.

The four diodes D_1, D_2, D_3, D_4 are arranged in series with only two diodes allowing electric current during each half cycle. For example, diodes D_1 and D_3 are considered as one pair which allows electric current during the positive half cycle whereas diodes D_2 and D_4 are considered as another pair which allows electric current during the negative half cycle of the input AC signal.

Working Principle of Bridge Rectifier

When input AC signal is applied across the bridge rectifier, during the positive half cycle diodes D_1

and D_3 are forward biased and allows electric current while the diodes D_2 and D_4 are reverse biased and blocks electric current. On the other hand, during the negative half cycle diodes D_2 and D_4 are forward biased and allow electric current while diodes D_1 and D_3 are reverse biased and blocks electric current.

During the positive half cycle, the terminal A becomes positive while the terminal B becomes negative. This causes the diodes D_1 and D_3 forward biased and at the same time, it causes the diodes D_2 and D_4 reverse biased.

The current flow direction during the positive half cycle is shown in the figure below (I.e. A to D to C to B).

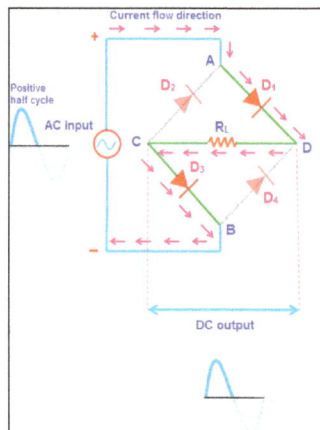

Bridge rectifier during positive half cycle

During the negative half cycle, the terminal B becomes positive while the terminal A becomes negative. This causes the diodes D_2 and D_4 forward biased and at the same time, it causes the diodes D_1 and D_3 reverse biased.

The current flow direction during negative half cycle is shown in the figure below (I.e. B to D to C to A).

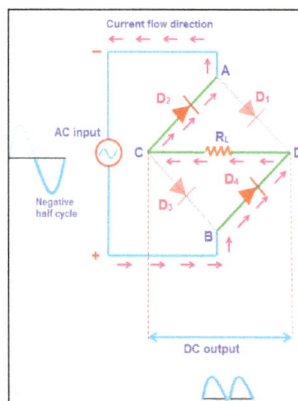

Bridge rectifier during negative half cycle

From the above two figures (A and B), we can observe that the direction of current flow across load resistor R_L is same during the positive half cycle and negative half cycle. Therefore, the polarity of the output DC signal is same for both positive and negative half cycles. The output DC signal polarity may be either completely positive or negative. In our case, it is completely positive. If the direction of diodes is reversed then we get a complete negative DC voltage.

Thus, a bridge rectifier allows electric current during both positive and negative half cycles of the input AC signal.

The output waveforms of the bridge rectifier is shown in the below figure.

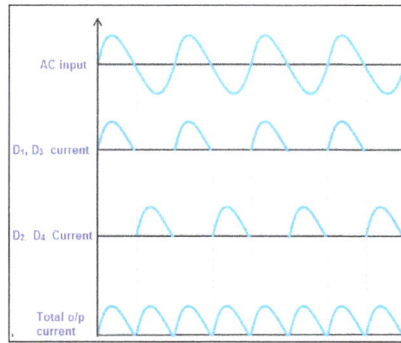

Characteristics of Bridge Rectifier

Peak Inverse Voltage (PIV)

The maximum voltage a diode can withstand in the reverse bias condition is called Peak Inverse Voltage (PIV).

Or

The maximum voltage that the non-conducting diode can withstand is called Peak Inverse Voltage (PIV).

During the positive half cycle, the diodes D_1 and D_3 are in the conducting state while the diodes D_2 and D_4 are in the non-conducting state. On the other hand, during the negative half cycle, the diodes D_2 and D_4 are in the conducting state while the diodes D_1 and D_3 are in the non-conducting state.

The Peak Inverse Voltage (PIV) for a bridge rectifier is given by:

$$PIV = V_{Smax}$$

Ripple Factor

The smoothness of the output DC signal is measured by using a factor known as ripple factor. The output DC signal with very fewer ripples is considered as the smooth DC signal while the output DC signal with high ripples is considered as the high pulsating DC signal.

Ripple factor is mathematically defined as the ratio of ripple voltage to the pure DC voltage. The ripple factor for a bridge rectifier is given by:

$$\gamma = \sqrt{\left(\frac{V_{rms}}{V_{DC}}\right)^2 - 1}$$

The ripple factor of the bridge rectifier is 0.48 which is same as the center tapped full wave rectifier.

Rectifier Efficiency

The rectifier efficiency determines how efficiently the rectifier converts Alternating Current (AC) into Direct Current (DC).

High rectifier efficiency indicates a most reliable rectifier while the low rectifier efficiency indicates a poor rectifier.

Rectifier efficiency is defined as the ratio of the DC output power to the AC input power.

$$\eta = \frac{\text{DC output power}}{\text{AC input power}} = \frac{P_{DC}}{P_{AC}}$$

The maximum rectifier efficiency of a bridge rectifier is 81.2% which is same as the center tapped full wave rectifier.

Advantages of Bridge Rectifier

1. Low ripples in the output DC signal: The DC output signal of the bridge rectifier is smoother than the half wave rectifier. In other words, the bridge rectifier has fewer ripples as compared to the half wave rectifier. However, the ripple factor of the bridge rectifier is same as the center tapped full wave rectifier.

2. High rectifier efficiency: The rectifier efficiency of the bridge rectifier is very high as compared to the half wave rectifier. However, the rectifier efficiency of bridge rectifier and center tapped full wave rectifier is same.

3. Low power loss: In half wave rectifier only one half cycle of the input AC signal is allowed and the remaining half cycle of the input AC signal is blocked. As a result, nearly half of the applied input power is wasted.

However, in the bridge rectifier, the electric current is allowed during both positive and negative half cycles of the input AC signal. So the output DC power is almost equal to the input AC power.

Disadvantages of bridge Rectifier

1. Bridge rectifier circuit looks very complex: In a half wave rectifier, only a single diode is used whereas in a center tapped full wave rectifier two diodes are used. But in the bridge rectifier, we use four diodes for the circuit operation. So the bridge rectifier circuit looks more complex than the half wave rectifier and center tapped full wave rectifier.

2. More power loss as compared to the Center tapped full wave rectifier: In electronic circuits, the more diodes we use the more voltage drop will occur. The power loss in bridge rectifier is almost equal to the center tapped full wave rectifier. However, in a bridge rectifier, the voltage drop is slightly high as compared to the center tapped full wave rectifier. This is due to two additional diodes (total four diodes).

In center tapped full wave rectifier, only one diode conducts during each half cycle. So the voltage drop in the circuit is 0.7 volts. But in the bridge rectifier, two diodes which are connected in series conduct during each half cycle. So the voltage drop occurs due to two diodes which is equal to 1.4 volts (0.7 + 0.7 = 1.4 volts). However, the power loss due to this voltage drop is very small.

AC to AC Converter

AC to AC Converter

AC to AC Converters are used to convert the alternating current waveforms of specific magnitude and specific frequency into alternating current waveforms with different magnitude and different frequency, as the signal remains AC, hence this converter is termed as AC to AC converter. To operate a few devices and machines we require some specific voltage with specific frequency which can be obtained using AC to AC converters. By regulating AC power using the AC to AC converter we can regulate the speed of induction motors. There are different types of AC to AC converters which are classified based on different criteria.

Cycloconverter

A cycloconverter refers to a frequency changer that can to change AC power from one frequency to AC power at another frequency. This process is known as AC-AC conversion. It is mainly used in electric traction, AC motors having variable speed and induction heating.

A cycloconverter can achieve frequency conversion in one stage and ensures that voltage and the frequencies are controllable. In addition, the need to use commutation circuits is not necessary because it utilizes natural commutation. Power transfer within a cycloconverter occurs in two directions (bidirectional).

A major problem with cycloconverters is that when it is operating at small currents, there are inefficiencies created with firing delay. Furthermore, operations are only smooth at frequencies that are not equal half frequency input values. This is true because a cycloconverter is an AC- AC converter that is phase controlled. Therefore, for it to give the required AC output voltage, it has to do a selection of the voltage input segments by applying line (natural) commutation. This explains why the output frequency is lower than the frequency input.

Harmonics in a cycloconverter are mainly affected by methods of control, overlap effect, the number of pulses in a given cycle, operation mode and mode of conduction.

There are two types of cycloconverters:

- Step Up cycloconverter – These types use natural commutation and give an output at higher frequency than that of the input.

- Step Down cycloconverter – This type uses forced commutation and results in an output with a frequency lower than that of the input.

- Cycloconverters are further classified into three categories:

- Single phase to single-phase – This type of cycloconverter has two full wave converters connected back to back. If one converter is operating the other one is disabled, no current passes through it.

- Three-phase to single-phase – This cycloconverter operates in four quadrants that is (+V, +I) and (–V, –I) being the rectification modes and (+V, –I) and (–V, +I) being the inversion modes.

- Three-phase to three-phase – This type of cycloconverter is majorly used in AC machine systems that are operating on three phase induction and synchronous machines.

Basic Principle of Operation

The basic principle of operation of a cyclo-converter is explained with reference to an equivalent circuit shown in figure. Each two-quadrant converter (phase-controlled) is represented as an alternating voltage source, which corresponds to the fundamental voltage component obtained at its output terminals. The diodes connected in series with each voltage source, show the unidirectional conduction of each converter, whose output voltage can be either positive or negative, being a two-quadrant one, but the direction of current is in the direction as shown in the circuit, as only thyristors – unidirectional switching devices, are used in the two converters. Normally, the ripple content in the output voltage is neglected.

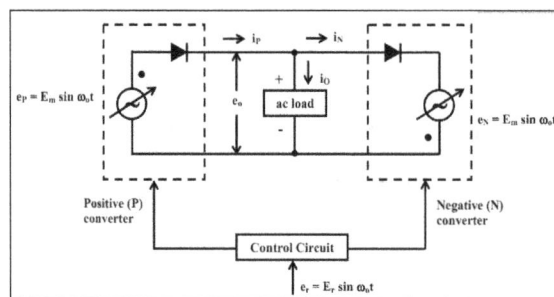

Equivalent circuit of cyclovonverter

The control principle used in an ideal cyclo-converter is to continuously modulate the firing angles of the individual converters, so that each produces the same sinusoidal (ac) voltage at its output terminals. Thus, the voltages of the two generators have the same amplitude, frequency and phase, and the voltage of the cyclo-converter is equal to the voltage of either of these generators. It is possible for the mean power to flow either 'to' or 'from' the output terminals, and the cyclo-converter is inherently capable of operation with loads of any phase angle – inductive or capacitive. Because of the uni-directional current carrying property of the individual converters, it is inherent that the positive half-cycle of load current must always be carried by the positive converter, and the negative half-cycle by the negative converter, regardless of the phase of the current with respect to the voltage. This means that each two-quadrant converter operates both in its rectifying (converting) and in its inverting region during the period of its associated half-cycle of current.

The output voltage and current waveforms, illustrating the operation of an ideal cycloconverter circuit with loads of various displacement angles, are shown in figure. The displacement angle of the load (current) 0° is. In this case, each converter carries the load current only, when it operates in its rectifying region, and it remains idle throughout the whole period in which its terminal voltage is in the inverting region of operation. In figure, the displacement angle of the load is 60° lagging. During the first 120° period of each half-cycle of load current, the associated converter operates in its rectifying region, and delivers power to the load. During the latter 60° period in the half-cycle, the associated converter operates in its inverting region, and under this condition, the load is regenerating power back into the cyclo-converter output terminals, and hence, into the ac system at the input side. These two are illustrative cases only. Any other case, say capacitive load, with the displacement angle as leading, the operation changes with inverting region in the first period of the half-cycle as per displacement angle, and the latter period operating in rectifying region.

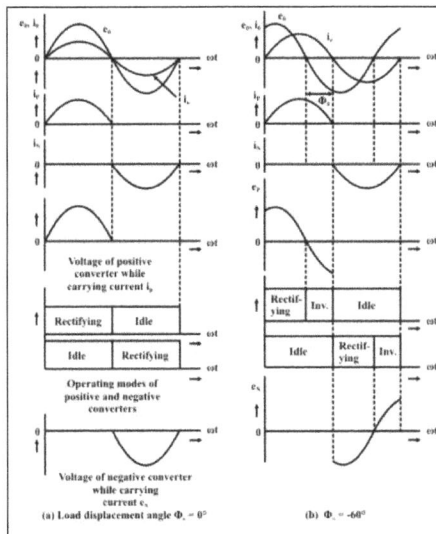

(a) Load displacement angle $\Phi_o = 0°$ (b) $\Phi_o = -60°$

Single-phase to Single-phase Cyclo-converter

The circuit of a single-phase to single-phase cyclo-converter is shown in figure. Two full-wave fully controlled bridge converter circuits, using four thyristors for each bridge, are connected in opposite direction (back to back), with both bridges being fed from ac supply (50 Hz). Bridge 1 (P – positive) supplies load current in the positive half of the output cycle, while bridge 2 (N – negative) supplies load current in the negative half. The two bridges should not conduct together as this will produce short-circuit at the input. In this case, two thyristors come in series with each voltage source. When the load current is positive, the firing pulses to the thyristors of bridge 2 are inhibited, while the thyristors of bridge 1 are triggered by giving pulses at their gates at that time. Similarly, when the load current is negative, the thyristors of bridge 2 are triggered by giving pulses at their gates, while the firing pulses to the thyristors of bridge 1 are inhibited at that time. This is the circulating-current free mode of operation. Thus, the firing angle control scheme must be such that only one converter conduct at a time, and the change over of firing pulses from one converter to the other, should be periodic according to the output frequency. However, the firing angles the thyristors in both converters should be the same to produce a symmetrical output.

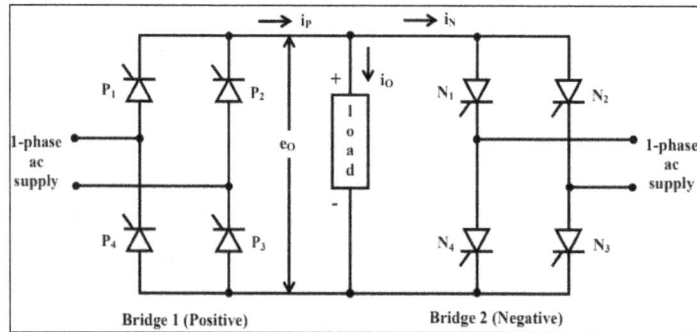

Single-phase to single-phase cycloconverter (using thyristor bridges)

When a cyclo-converter operates in the non-circulating current mode, the control scheme is complicated, if the load current is discontinuous. The control is somewhat simplified, if some amount of circulating current is allowed to flow between them. In this case, a circulating current limiting reactor is connected between the positive and negative converters, as is the case with dual converter, i.e. two fully controlled bridge converters connected back to back, in circulating-current mode. The readers are requested to refer to any standard text book. This circulating current by itself keeps both converters in virtually continuous conduction over the whole control range. This type of operation is termed as the circulating-current mode of operation. The operation of the cyclo-converter circuit with both purely resistive (R), and inductive (R-L) loads is explained.

1. Resistive (R) Load: For this load, the load current (instantaneous) goes to zero, as the input voltage at the end of each half cycle (both positive and negative) reaches zero (0). Thus, the conducting thyristor pair in one of the bridges turns off at that time, i.e. the thyristors undergo natural commutation. So, operation with discontinuous current takes place, as current flows in the load, only when the next thyristor pair in that bridge is triggered, or pulses are fed at respective gates. Taking first bridge 1 (positive), and assuming the top point of the ac supply as positive with the bottom point as negative in the positive half cycle of ac input, the oddnumbered thyristor pair, P_1 & P_3 is triggered after phase delay (α_1), such that current starts flowing through the load in this half cycle. In the next (negative) half cycle, the other thyristor pair (even-numbered), P_2 & P_4 in that bridge conducts, by triggering them after suitable phase delay from the start of zero-crossing. The current flows through the load in the same direction, with the output voltage also remaining positive. This process continues for one more half cycle (making a total of three) of input voltage$\left(f_1 = 50 \text{ Hz} \right)$. From three waveforms, one combined positive half cycle of output voltage is produced across the load resistance, with its frequency being one-third of input frequency$\left(f_2 = f_1 / 3 = 16\frac{2}{3} \text{Hz} \right)$.

The following points may be noted. The firing angle (α) of the converter is first decreased, in this case for second cycle only, and then again increased in the next (third) cycle, as shown in figure. This is, because only three cycles for each half cycle is used. If the output frequency needed is lower, the number of cycles is to be increased, with the firing angle decreasing for some cycles, and then again increasing in the subsequent cycles.

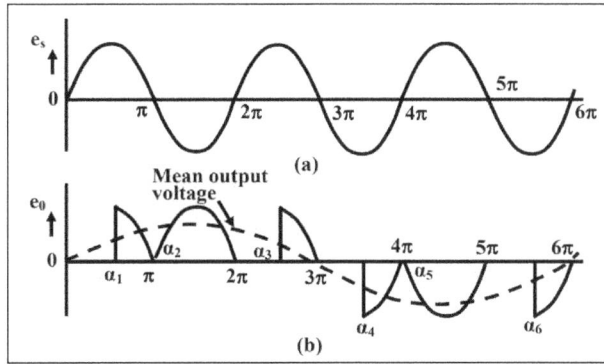

Input (a) and output (b) voltage waveforms of a cyclo-converter with an output frequency of $16\frac{2}{3}$ Hz for resistive (R) load

To obtain negative output voltage, in the next three half cycles of input voltage, bridge 2 is used. Following same logic, if the bottom point of the ac supply is taken as positive with the top point as negative in the negative half of ac input, the odd-numbered thyristor pair, N_1 & N_3 conducts, by triggering them after suitable phase delay from the zero-crossing. Similarly, the even-numbered thyristor pair, N_2 & N_4 conducts in the next half cycle. Both the output voltage and current are now negative. As in the previous case, the above process also continues for three consecutive half cycles of input voltage. From three waveforms, one combined negative half cycle of output voltage is produced, having same frequency. The pattern of firing angle – first decreasing and the increasing, is also followed in the negative half cycle. One positive half cycle, along with one negative half cycle, constitute one complete cycle of output (load) voltage waveform, its frequency being $16\frac{2}{3}$ Hz as stated earlier. The ripple frequency of the output voltage/current for single–phase full-wave converter is 100 Hz, i.e., double of the input frequency. It may be noted that the load (output) current is discontinuous, as also load (output) voltage. The supply (input) voltage is shown in figure. Only one of two thyristor bridges (positive or negative) conducts at a time, giving non-circulating current mode of operation in this circuit.

2. Inductive (R-L) Load: For this load, the load current may be continuous or discontinuous depending on the firing angle and load power factor. The load voltage and current waveforms are shown for continuous and discontinuous load current in figure and respectively.

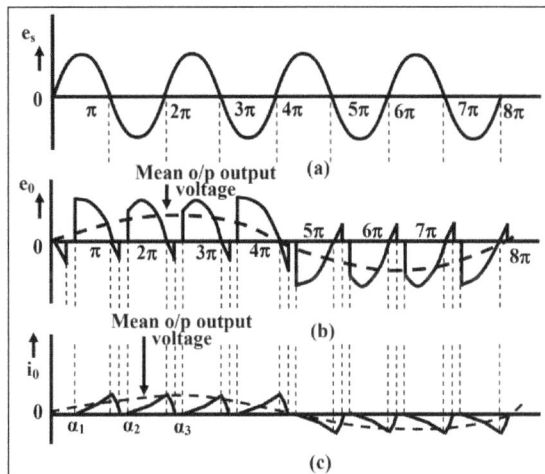

Input (a) and output (b) voltage, and current (c) waveforms for a cyclo-converter with discontinuous

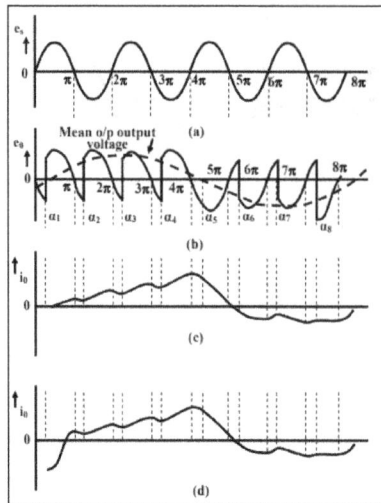

Input (a) and output (b) voltage, and current (c, d) waveforms for a cyclo-converter
with continuous load current

Discontinuous Load Current

The load current is discontinuous, as the inductance of the load is low. If the inductance is increased, the current will be continuous. Most of the points given earlier are applicable to this case, as described. To repeat, non-circulating mode of operation is used, i.e., only one of the bridges – 1 (positive), or 2 (negative), conducts at a time, but two bridges do not conduct at the same time, as this will result in a short circuit. Also, the ripple frequency in the voltage and current waveforms remains same at 100 Hz. The output frequency is one-fourth of input frequency (50 Hz), i.e., 12.5 Hz. So, for each half-cycle of output voltage waveform, four half cycles of input supply are required. Taking bridge 1, and assuming the top point of the ac supply as positive, in the positive half cycle of ac input, the odd-numbered thyristor pair, P_1 & P_3, is triggered after phase delay $(\theta = \omega = \alpha_1)$, such that current starts flowing the inductive load in this half cycle. But here, the current flows for about one complete half cycle, i.e., up to the angle, $(\pi + \alpha_1)$ or $(\pi + \alpha_2)$, whichever is higher, even after the input voltage has reversed, due to the high value of load inductance. In the next (negative) half cycle, the other thyristor pair (even-numbered), P_2 & P_4, is triggered at $(\pi + \alpha_2)$. At that time, reverse voltage is applied across each of the conducting thyristors, P_1/P_3, and the thyristors turn off. The current flows through the load in the same direction, with the output voltage also remaining positive. Also, the current flows for about one complete half cycle, i.e., up to the angle, $(\pi + \alpha_2)$ or $(\pi + \alpha_3)$, whichever is higher. This procedure continues for the next two half cycles, making a total of four positive half cycles. From these four waveforms, one combined positive half cycle of output voltage is produced across the inductive load. The firing angle (α) of the converter is first decreased, in this case for second half cycle only, kept nearly same in the third one, and finally increased in the last (fourth) one.

To obtain negative output voltage, in the next four half cycles of output voltage, bridge 2 is used. Following same logic, if the bottom point of the ac supply is taken as positive in the negative half of ac input, the odd-numbered thyristor pair, N_1 & N_3 conducts, by triggering them after phase delay $(\theta = 4 \cdot \pi + \alpha_1)$. The current flows now in the opposite (negative) direction through the inductive load, with the output voltage being also negative. The current flows for about one complete half cycle, i.e., up to the angle, $(5 \cdot \pi + \alpha_1)$ or $(5 \cdot \pi + \alpha_2)$, whichever is higher, as the load is inductive.

Similarly, the even-numbered thyristor pair, N_2 & N_4 conducts in the next half cycle, after they are triggered at $(5 \cdot \pi + \alpha_2)$. Both the conducting thyristors turn off, as reverse voltage is applied across each of them. Both the output voltage and current are now negative. Also, the current flows for about one complete half cycle, i.e. up to the angle, $(5 \cdot \pi + \alpha_2)$ or $(5 \cdot \pi + \alpha_3)$, whichever is higher. As in the previous case, the above process also continues for two more half cycles of input voltage, making a total of four. From these four waveforms, one combined negative half cycle of output voltage is produced with same output frequency of 12.5 Hz. The pattern of firing angle – first decreasing and then increasing, is also followed in the negative half cycle. It may be observed that the load (output) current is continuous, as also load (output) voltage. The load (output) current is redrawn, under steady state condition, while the supply (input) voltage is shown in figure. One positive half cycle, along with one negative half cycle, constitute one complete cycle of output (load) voltage waveform.

Advantages and Disadvantages of Cycloconverter

Advantages

1. In a cycloconverter, ac power at one frequency is converted directly to a lower frequency in a single conversion stage.

2. Cycloconverter functions by means of phase commutation, without auxiliary forced commutation circuits. The power circuit is more compact, eliminating circuit losses associated with forced commutation.

3. Cycloconverter is inherently capable of power transfer in either direction between source and load. It can supply power to loads at any power factor, and is also capable of regeneration over the complete speed range, down to standstill. This feature makes it preferable for large reversing drives requiring rapid acceleration and deceleration, thus suited for metal rolling application.

4. Commutation failure causes a short circuit of ac supply. But, if an individual fuse blows off, a complete shutdown is not necessary, and cyclo-converter continues to function with somewhat distorted waveforms. A balanced load is presented to the ac supply with unbalanced output conditions.

5. Cycloconverter delivers a high quality sinusoidal waveform at low output frequencies, since it is fabricated from a large number of segments of the supply waveform. This is often preferable for very low speed applications.

6. Cycloconverter is extremely attractive for large power, low speed drives.

Disadvantages

1. Large number of thyristors is required in a cycloconverter, and its control circuitry becomes more complex. It is not justified to use it for small installations, but is economical for units above 20 kVA.

2. For reasonable power output and efficiency, the output frequency is limited to one-third of the input frequency.

3. The power factor is low particularly at reduced output voltages, as phase control is used with high firing delay angle.

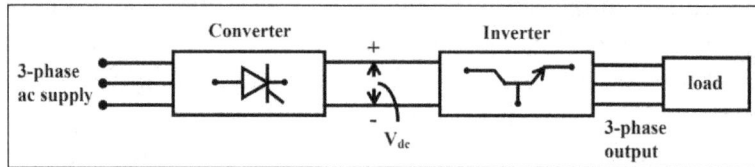

DC Link Converter

The cycloconverter is normally compared with dc link converter, where two power controllers, first one for converting from ac input at line frequency to dc output, and the second one as inverter to obtain ac output at any frequency from the above dc input fed to it. The thyristors, or switching devices of transistor family, which are termed as self-commutated ones, usually the former, which in this case is naturally commutated, are used in controlled converters (rectifiers). The diodes, whose cost is low, are used in uncontrolled ones. But now-a-days, switching devices of transistor family are used in inverters, though thyristors using force commutation are also used. A diode, connected back to back with the switching device, may be a power transistor (BJT), is needed for each device. The number of switching devices in dc link converter depends upon the number of phases used at both input and output. The number of devices, such as thyristors, used in cyclo-converters depends on the types of connection, and also the number of phases at both input and output. It may be noted that all features of a cycloconverter may not be available in a dc link converter. Similarly, certain features, like Pulse Width Modulation (PWM) techniques as used in inverters and also converters, to reduce the harmonics in voltage waveforms, are not applied in cycloconverters.

Advantages and Disadvantages of DC Link Converter

Advantages

1. The output frequency can be varied from zero to rated value, with the upper frequency limit, being decided by the turn-off time of the switching devices, which is quite low due to the use of transistors in recent time.

2. The control circuit here is simpler, as compared to that used in cyclo-converter.

3. It has high input power factor, if diode rectifier is used in the first stage. If phase-controlled thyristor converter is used, power factor depends upon phase angle delay.

4. It is suitable for higher frequencies.

Disadvantages

1. The conversion is in two stages, using two power controllers – one as converter and other as inverter, as stated earlier.

2. Forced commutation is required for the inverter, if thyristors are used, even though phase control is used in converter, where natural commutation takes place.

3. The feature of regeneration is somewhat difficult, and also is involved to incorporate in a dc link converter.

4. The output waveform of the inverter is normally a stepped one, which may cause nonuniform rotation of an ac motor at very low frequencies (< 10 Hz). The distorted waveform also causes system instability at low frequencies. This can be reduced by using PWM technique.

Matrix Converters

A matrix converter is defined as a converter with a single stage of conversion. It utilizes bidirectional controlled switch to achieve automatic conversion of power from AC to AC. It provides an alternative to PWM voltage rectifier (double sided).

Matrix converters are characterized by sinusoidal waveforms that show the input and output switching frequencies. The bidirectional switches make it possible to have a controllable power factor input. In addition, the lack of DC links ensures it has a compact design. The downside to matrix converters is that they lack bilateral switches that are fully controlled and able to operate at high frequencies. Its voltage ratio that is output to input voltage is limited.

There are three methods of matrix converter control:

1. Space vector modulation,

2. Pulse width modulation,

3. Venturi - analysis of function transfer.

The Matrix Converter Circuit

The diagram given below shows a single-phase matrix converter.

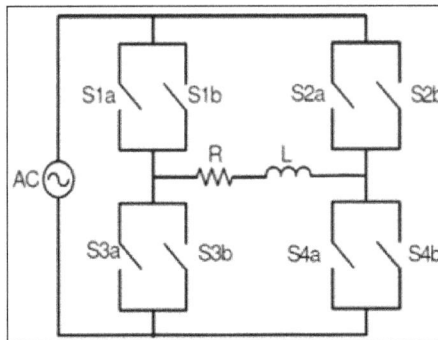

It contains four bi-directional switches with each switch having the ability to conduct in both forward blocking and reverse voltage.

Space Vector Modulation (SVM)

SVM refers to a method of algorithm used to control the PWM. It creates AC waveforms that drive AC motors at various speeds. In the case of a three-phase inverter having DC supply power, its three main legs at the output are connected to a 3-phase motor.

The switches are under control to ensure that no two switches in the same leg are ON at the same time. Simultaneous ON states could result in the DC supply shorting. This leads to eight switching vectors where two are zero and six are active vectors for switching.

DC to DC Converter

Chopper is a static power electronics device which converts fixed DC voltage/power to variable DC voltage or power. It is nothing but a high speed switch which connects and disconnects the load from source at a high rate to get variable or chopped voltage at the output.

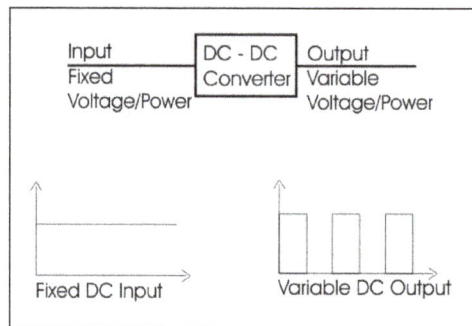

Chopper can increase or decrease the DC voltage level at its opposite side. So, chopper serves the same purpose in DC circuit transfers in case of ac circuit. So it is also known as DC transformer.

Devices used in Chopper

- Low power application: GTO, IGBT, Power BJT, Power MOSFET etc.

- High power application: Thyristor or SCR.

These devices are represented as a switch in a dotted box for simplicity. When it is closed current can flow in the direction of arrow only.

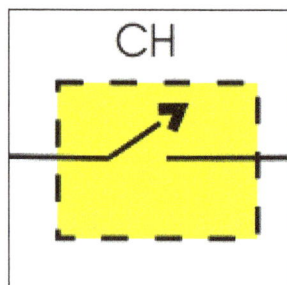

Step Down Chopper

Step down chopper as Buck converted is used to reduce the i/p voltage level at the output side. Circuit diagram of a step down chopper is shown in the adjacent figure.

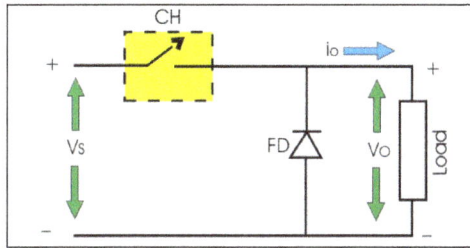

When CH is turned ON, V_s directly appears across the load as shown in figure.

So $V_o = V_s$.

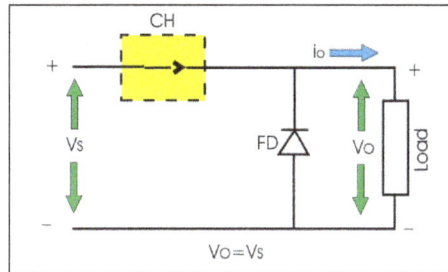

When CH is turned off, V_s is disconnected from the load. So output voltage $V_o = 0$.

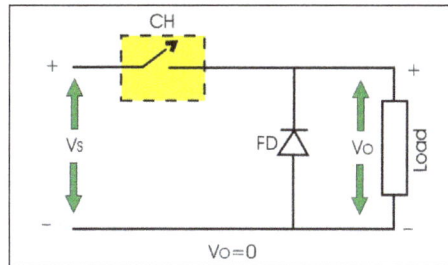

The voltage waveform of step down chopper is shown below:

- $T_{ON} \rightarrow$ It is the interval in which chopper is in ON state.

- $T_{OFF} \rightarrow$ It is the interval in which chopper is in OFF state.

- $V_S \rightarrow$ Source or input voltage.

- $V_o \rightarrow$ Output or load voltage.

- $T \rightarrow$ Chopping period $= T_{ON} + T_{OFF}$.

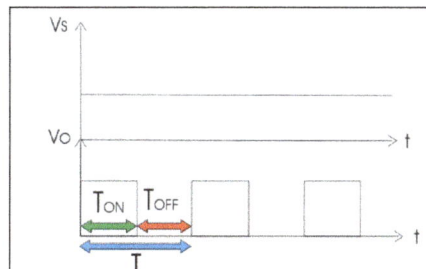

Operation of Step down Chopper with Resistive Load

When CH is ON, $V_o = V_s$.

When CH is OFF, $V_o = 0$.

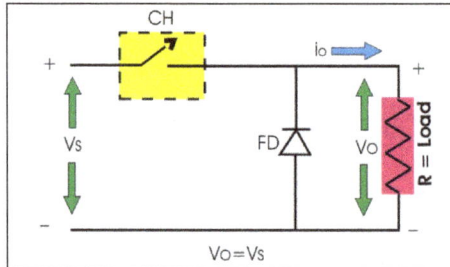

$$\text{Average output voltage} = V_O = \frac{1}{T}\int_0^{T_{ON}} V_s dt = \frac{V_s T_{ON}}{T} = DV_s$$

Where, D is duty cycle = T_{ON}/T.

T_{ON} can be varied from 0 to T, so $0 \le D \le 1$. Hence output voltage V_o can be varied from 0 to V_s.

$$\text{RMS output voltage} = V_{or} = \sqrt{\frac{1}{T}\int_0^{T_{ON}} V_s^2 dt} = V_s\sqrt{\frac{T_{ON}}{T}} = \sqrt{D}V_s$$

Therefore,

$$\text{Effective input resistance} = R_i = \frac{V_s}{T_{savg}} = \frac{V_s}{DV_s/R} = \frac{R}{D}.$$

So, we can conclude that output voltage is always less than the input voltage and hence the name step down chopper is justified. The output voltage and current waveform of step down chopper with resistive load is shown in the figure.

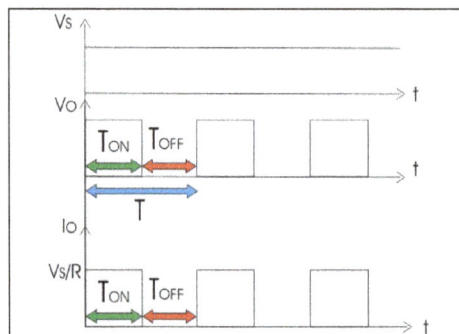

Operation of Step down Chopper with Inductive Load

When CH is ON, $V_o = V_s$.

When CH is OFF, $V_o = 0$.

During ON time of Chopper

$$V_s = V_L + V_o \Rightarrow V_L = V_s - V_o \Rightarrow L\frac{di}{dt} = V_s - V_o \Rightarrow L\frac{\Delta I}{T_{ON}} = V_s - V_o$$

Therefore, peak to peak load current:

$$\Delta I = \frac{V_s - V_o}{L} T_{ON}$$

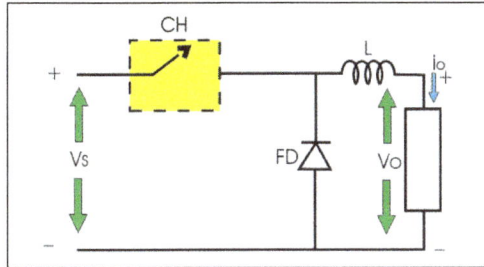

During OFF Time of Chopper

If inductance value of L is very large, so load current will be continuous in nature. When CH is OFF inductor reverses its polarity and discharges. This current freewheels through diode FD.

Therefore,

$$L\frac{di}{dt} = V_o$$

$$L\frac{\Delta I}{T_{OFF}} = V_o \Rightarrow \Delta I = V_o \frac{T_{OFF}}{L}$$

By equating $\Delta I = \frac{V_s - V_o}{L} T_{ON}$ and $L\frac{\Delta I}{T_{OFF}} = V_o \Rightarrow \Delta I = V_o \frac{T_{OFF}}{L}$

$$\frac{V_s - V_o}{L} T_{ON} = \frac{V_o}{L} T_{OFF}$$

$$\frac{V_s - V_o}{V_o} = \frac{T_{OFF}}{T_{ON}}$$

$$\frac{V_s}{V_o} = \frac{T_{ON} - T_{OFF}}{T_{ON}}$$

$$V_o = \frac{T_{ON}}{T} V_S, = DV_s$$

So, from $\Delta I = \frac{V_s - V_o}{L} T_{ON}$ we get,

$$\Delta I = \frac{V_s - DV_s}{L} DT \quad \left[Since, \ D = \frac{T_{ON}}{T} \right]$$

$$= \frac{V_s(1-D)D}{Lf} \quad \left[f = \frac{1}{T} = Chopping \ Frequency \right]$$

The output voltage and current waveform of step down chopper with inductive load is shown in the figure.

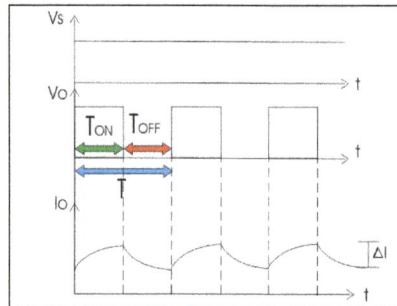

Step up Chopper or Boost Converter

Step up chopper or boost converter is used to increase the input voltage level of its output side. Its circuit diagram and waveforms are shown below in figure.

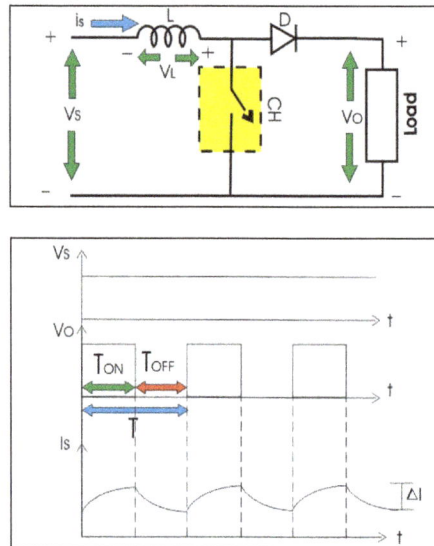

Operation of Step up Chopper

When CH is ON it short circuits the load. Hence output voltage during T_{ON} is zero. During this period inductor gets charged. So, $V_S = V_L$:

$$L \frac{di}{dt} = V_s \Rightarrow \frac{\Delta I}{T_{ON}} = \frac{V_s}{L} \Rightarrow \Delta I = \frac{V_s}{L} T_{ON}$$

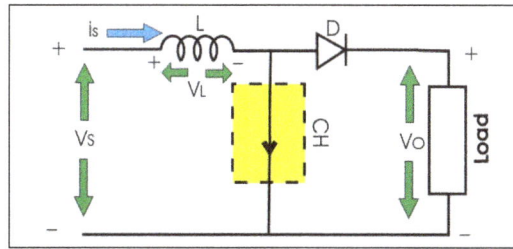

Where, ΔI is the peak to peak inductor current.

When CH is OFF inductor L discharges through the load. So, we will get summation of both source voltage V_s and inductor Voltage V_L as output voltage, i.e.

$$V_o = V_s + V_L \Rightarrow V_L = V_o - V_s \Rightarrow L\frac{di}{dt} V_o - V_s$$

$$\Rightarrow L\frac{\Delta I}{T_{OFF}} = V_o - V_s \Rightarrow \Delta I = \frac{V_o - V_s}{L} T_{OFF}$$

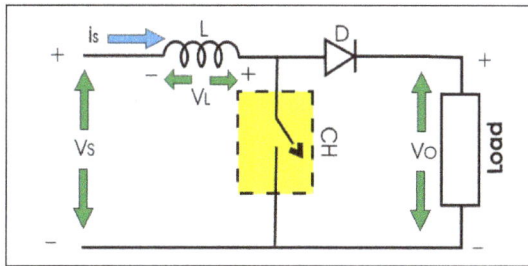

$$\text{Now, by equating } L\frac{di}{dt} = V_s \Rightarrow \frac{\Delta I}{T_{ON}} = \frac{V_s}{L} \Rightarrow \Delta I = \frac{V_s}{L} T_{ON} \text{ and} \Rightarrow L\frac{\Delta I}{T_{OFF}} = V_o - V_s \Rightarrow \Delta I = \frac{V_o - V_s}{L} T_{OFF},$$

$$\frac{V_s}{L} T_{ON} = \frac{V_o - V_s}{L} T_{OFF} \Rightarrow V_s (T_{ON} + T_{OFF}) = V_o T_{OFF}$$

$$\Rightarrow V_o = \frac{TV_s}{T_{OFF}} = \frac{V_s}{(T - T_{ON})/T}$$

Average output voltage, $V_o = \dfrac{V_s}{1-D}$

As we can vary TON from 0 to T, so $0 \leq D \leq 1$. Hence V_O can be varied from V_S to ∞. It is clear that output voltage is always greater than the input voltage and hence it boost up or increase the voltage level.

Buck-Boost Converter or Step Up Step Down Converter

With the help of Buck-Boost converter we can increase or decrease the input voltage level at its output side as per our requirement. The circuit diagram of this converter is shown in the figure.

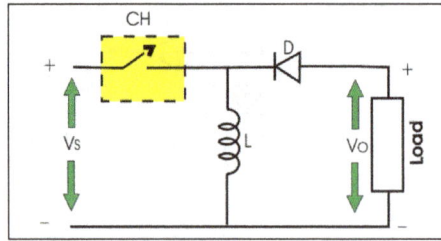

Operation of Buck-Boost Converter

When CH is ON source voltage will be applied across inductor L and it will be charged.

So, $V_L = V_s$

$$L\frac{di}{dt} = V_s \Rightarrow \Delta I = \frac{V_s}{L}T_{ON}$$

$$\Rightarrow \Delta I = \frac{V_s}{L}T\frac{T_{ON}}{T}$$

$$\Delta I = \frac{DV_s}{Lf} \quad \left[Since, \frac{T_{ON}}{T} = D \ and \ \frac{1}{T} = f = Chopping \ Frequencey \right]$$

When chopper is OFF inductor L reverses its polarity and discharges through load and diode,

So, V_o - V_L

$$L\frac{di}{dt} = -V_o \Rightarrow L\frac{\Delta I}{T_{OFF}} = -V_o \Rightarrow \left| Delta = -\frac{V_o}{L}T_{OFF} \right.$$

By evaluating $\Delta I = \frac{DV_s}{Lf}$ $\left[Since, \frac{T_{ON}}{T} = D \ and \ \frac{1}{T} = f = Chopping \ Frequencey \right]$ and

$$L\frac{di}{dt} = -V_o \Rightarrow L\frac{\Delta I}{T_{OFF}} = -V_o \Rightarrow \left| Delta = -\frac{V_o}{L}T_{OFF} \right. \quad \text{we get:}$$

$$\frac{DV_s}{fL} = -\frac{V_o}{L}T_{OFF} \Rightarrow DV_s = -V_o T_{OFF} f$$

$$DV_s = -V_o\frac{T - T_{ON}}{T} = -V_o\left(1 - \frac{T_{ON}}{T}\right) \Rightarrow V_o = -\frac{DV_s}{1-D} \quad \left[Since, D = \frac{T_{ON}}{T} = \frac{T - T_{OFF}}{T} \right]$$

Taking magnitude we get:

$$V_o = \frac{DV_s}{1-D}$$

D can be varied from 0 to one.

- When, $D = 0$; $V_o = 0$

- When D = 0.5, V_o = VS

- When, D = 1, V_o = ∞

Hence, in the interval $0 \leq D \leq 0.5$, output voltage varies in the range $0 \leq V_o \leq V_S$ and we get step down or Buck operation.

Whereas, in the interval $0.5 \leq D \leq 1$, output voltage varies in the range $V_S \leq V_O \leq \infty$ and we get step up or Boost operation.

According to Direction of Output Voltage and Current

Semiconductors devices used in chopper circuit are unidirectional. But arranging the devices in proper way we can get output voltage as well as output current from chopper in our required direction. So, on the basis of this features chopper can be categorized as follows:

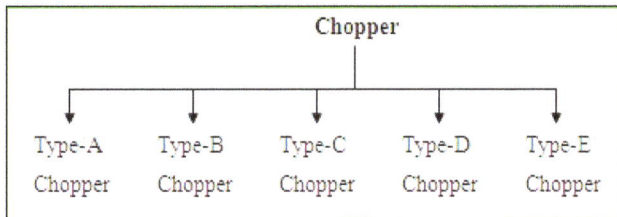

Before detailed analysis some basic idea regarding $V_o - I_o$ quadrant is required here.

The directions of I_o and V_o marked in the figure-1 is taken as positive direction.

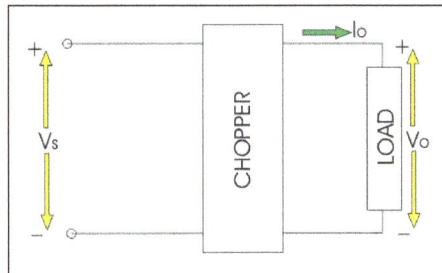

If output voltage (V_o) and output current (I_o) follows the direction as marked in figures then the chopper operation will be restricted in the first quadrant of $V_o - I_o$ plane. This type of operation is also known as forward motoring.

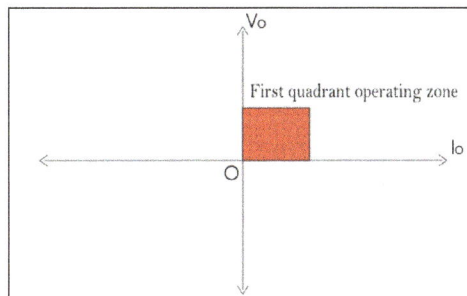

When output voltage (V_o) follows the marked direction in figure but current flows in the opposite direction then V_o is taken positive but I_o as negative. Hence the chopper operates in the second quadrant of $V_o - I_o$ plane. This type of operation is also known as forward braking.

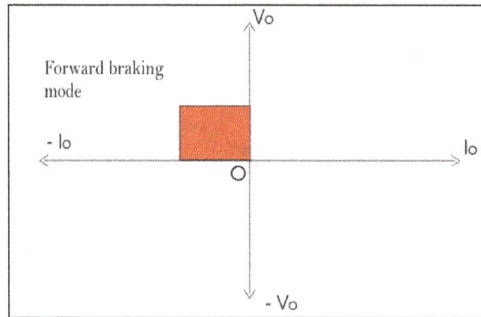

It may also happen that both output voltage and current is opposite to the marked direction in figure – 1. In this case both V_o and I_o are taken as negative. Hence chopper operation is restricted in third quadrant of V_o-I_o plane. This operation is called reverse motoring.

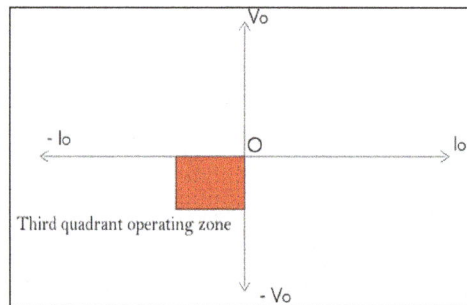

If output voltage is opposite to the marked direction in figure then it is taken as negative. But output current follows the direction as marked in figure and considered as positive. Hence chopper operates in 4th quadrant of $V_o - I_o$ plane. This mode of operation is called reverse braking.

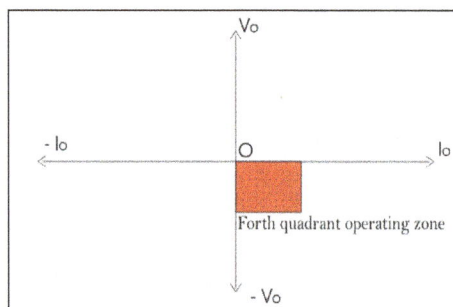

Now we can proceed to detailed analysis of different types of chopper. Some choppers operate in a single quadrant only, which are called single quadrant chopper. Some choppers operate in two quadrant also which are known as two quadrant chopper. It is also possible that a chopper operates in all the quadrants, which are known as 4-quadrant chopper.

Type-A Chopper

It is a single quadrant chopper whose operation is restricted in first quadrant of $V_o - I_o$ plane. The circuit diagram is shown in the figure.

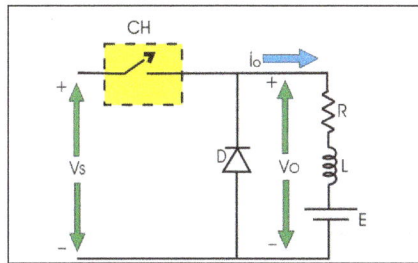

When CH is ON both V_o and I_o follows the direction as marked in the figures. So, both are taken as positive hence load power is positive which means power is delivered from source to land.

When CH is OFF current freewheels through diode. Hence V_o is zero and I_o is positive.

In type-A chopper it is seen that average value of V_o and I_o is always positive. This is also called step down chopper as average value of V_o is less than the input voltage. This type of chopper is suitable for motoring operation.

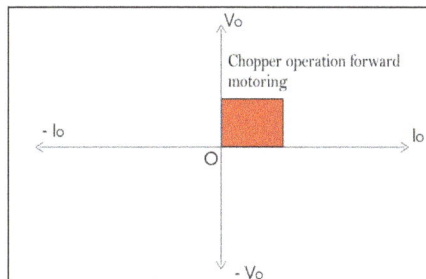

Type-B Chopper

This is also a single quadrant chopper operating in second quadrant of $V_o - I_o$ plane. The circuit diagram is shown in the following figure.

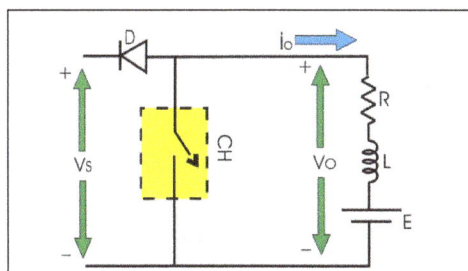

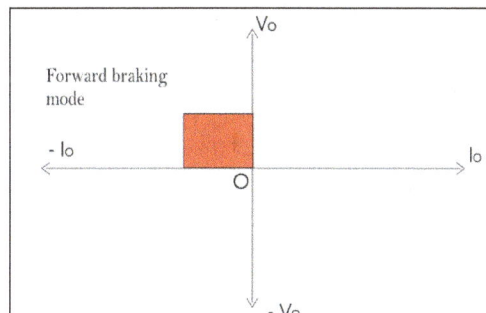

It is interesting to note that load must have a DC voltage source E for this kind of operation.

When CH is ON V_o is zero but current flows in the opposite direction as marked in figure. When chopper is OFF.

Which exceeds the source voltage V_S. So current flows through diode D and treated as negative.

Hence current I_o is always negative here but V_o is positive (sometimes zero). So, power flows from load to source and operation of type-B chopper is restricted in second quadrant of $V_o - I_o$ plane. This type of chopper is suitable for forward braking operation.

Type-C Chopper

This is a two quadrant chopper whose operation is bounded between first and second quadrant of $V_o - I_o$ plane. This type of chopper obtained by connecting type-A and type-B chopper in parallel as shown in the figure.

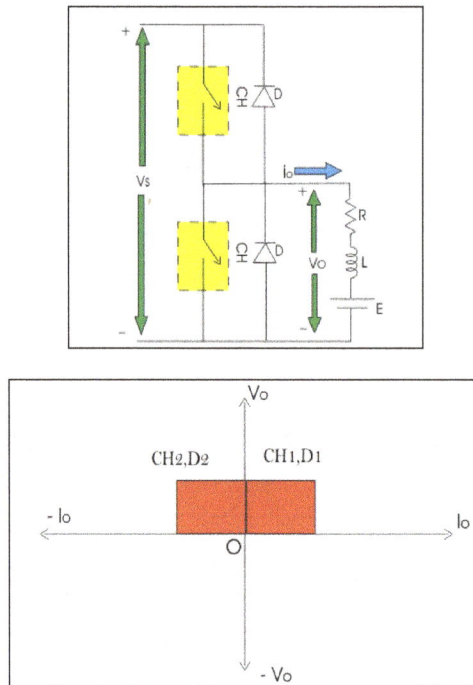

When CH_1 is ON current flows through abcdefa and inductor L will be charged. Hence output voltage V_o and current I_o both will be positive. When CH_1 is OFF, induction will discharge through D_1 and current I_o will flow through same direction with zero output voltage. So, we can see the operation of CH_1 is nothing but the operation of type-A chopper by which we can operate a chopper in the first quadrant.

When CH_2 is ON, output voltage V_o will be zero but output current I_o will flow in opposite direction of current shown in the figure and inductor will be charged up. When CH_2 is OFF Output voltage.

$$V_O = \left[E + L\frac{di}{dt} - IR \right]$$

Which exceeds the value of source voltage V_s. So current flows through diode D_2 and treated as negative. Hence output voltage V_o is always positive and output current I_o is always negative here. We can see operation of CH_2 is nothing but operation of type-B chopper by which we can operate the chopper in the second quadrant.

We can conclude that the operation of type-C chopper is the combined operation of type-A and type-B chopper. This type of chopper is suitable for both forward motoring and forward braking operation.

References

- What-is-an-inverter: sunpower-uk.com, Retrieved 9 May, 2019

- Rectifier-whatisrectifier, electronic-devices-and-circuits: physics-and-radio-electronics.com, Retrieved 19 February, 2019

- Half-wave-rectifiers: electrical4u.com, Retrieved 29 July, 2019

- Diode: electronics-tutorials.ws, Retrieved 1 April, 2019

- Bridgerectifier, rectifier, electronic-devices-and-circuits: physics-and-radio-electronics.com, Retrieved 11 April, 2019

- Simple-ac-to-ac-converter-circuit: edgefx.in, Retrieved 21 March, 2019

- Power-electronics-cycloconverters, power-electronics: tutorialspoint.com, Retrieved 20 August, 2019

- Power-electronics-matrix-converters, power-electronics: tutorialspoint.com, Retrieved 10 Janaury, 2019

- Chopper-dc-to-dc-converter: electrical4u.com, Retrieved 12 June, 2019

5

Power Generation and Distribution

The process through which usable electric current is distributed and produced is known as power generation. It can be generated through various sources such as solar energy, wind energy and geothermal energy. All the diverse principles of power generation and distribution have been carefully analyzed in this chapter.

Power Generation

Power generation is the process of producing and distributing usable electric current. Beyond utility providers, power generation plays an important role in manufacturing and industrial operations.

There are three main types of non-utility power generation:

- Freestanding generators use kerosene, gasoline, or other fossil fuels to trigger combustion that is converted into a usable power supply.

- Electrochemical power plants harness chemical reactions to supply their buildings and equipment.

- Energy from renewable resources like geothermal, solar, and wind power can be gathered and used to provide electrical power.

Power generation is required where the mains are either unavailable or insufficient. Typical applications in construction include cranes, conveyors, concrete mixing/pumping, pile driving, substation outages and the supply to hand held power tools.

A generator is a piece of equipment that basically converts diesel fuel into electrical power.

There are three types of generator which are stand alone, island and parallel.

- Stand-alone – Is not connected with other generators or the utility. As an isolated unit it supplies power to all connected load. Examples: emergency generators, aggregates or portable diesel generator sets.

- Island – In island operation, a generator is connected with other generators but not with the utility. As an isolated system the generators supply all power to the connected load. Examples: power systems on ships, on offshore platforms or in the desert.

- Parallel – Two generators working in parallel (load sharing) are used mainly for events where power loss would be costly and inconvenient. If for one reason one generator fails, the other generator will take over the load without any power interruption. Each generator must be capable of handling the full load alone.

Solar Power Generation

Solar power is the conversion of sun radiation into electricity through the use of solar photovoltaic cells. This conversion takes place in the solar cell by photovoltaic effect. As said by many experts that the amount of solar energy reaching the earth is more than 10000 times the current energy consumption by man.

Also, the power created by solar is sufficient for one year for the entire planet, if we could convert the 100 percent of the solar energy into electricity in one hour.

Solar power

There are several applications that use solar power, here is the information on the generation of electricity through PV cells. The solar power generation is the most efficient route for power generation because it takes a minimum number of steps (for producing electricity) than that of other generation methods.

There are two ways of converting sunlight into electricity. In one method, solar energy is used simply as a source of heat. This heat is further used to produce the steam, which drives the steam turbine. This method of power generation is called solar thermal power generation.

In the second method, solar energy is directly converted into electricity using PV (or solar) cells as mentioned above. The PV cell is made with silicon semiconductor material.

Some of the factors for choosing the solar power generation are listed below:

- Solar energy is available freely and conveniently in nature and it needs no mains supply.

- Solar generation plant can be installed in a few months while the conventional power plants take several years to build an electricity generation plant.

- Solar power is clean energy as it produces no air or water pollution. Also, there are no moving parts to create noise pollution. Unlike fossil fuels, no toxic emissions are released into the atmosphere during solar energy power generation.

- Solar power has less running cost that means once the capital investment is made, there is no need for continues purchase of fossil fuels as the solar energy is effectively free in nature.

Process of Solar Power Generation from Photovoltaic Cell

A PV cell (can be called as a solar cell) is a semiconductor device that converts the sunlight energy into electricity without going through any energy conversion steps.

This conversion takes place by photovoltaic effect and hence they are called Photovoltaic (PV) cells. It generates voltage and current at its terminals when sunlight incident on it.

PV cell

The way and the amount of power generated by a solar cell depend on the sunlight falling on it. This also includes some factors such as intensity of light, angle at which the light falls on it and area of the cell.

The more is the power generated, if higher is the light intensity. If the area of the cell is more, the power generated is also more. And the optimum power is generated by it when light falling is perpendicular to the front side of the cell.

The solar cells are made with silicon semiconductor material and is treated with phosphorous and boron to make a thin silicon wafer. The wafer layers are then aligned together to make the solar cells, once they are doped.

Irrespective of the technology and material used, every solar cell has two terminals (positive and negative terminals) so as to take the electric current from it. Typically, a solar cell consists of front contact at the top, PN junction in the middle and back contact at the bottom.

Basically, the sunlight consists of bundles of photons, where each photon has a finite amount of energy. To generate the electricity from a solar cell, these photons must be absorbed by it. The energy of the photon and also the band-gap energy of semiconductor material decide the absorption of a photon.

Here is the term Electron-volt (eV) which is the unit of energy that expresses the photon energy and the band-gap energy of a semiconductor material.

The semiconductor material of the solar panel absorbs the photons in the sunlight. Due to this, electron-hole pairs are generated at the junction. When the solar cell is connected to the load, electrons and holes at the junction are separated from each other where the electrons are collected at the negative terminal and holes at positive terminal.

Thus the electric potential is built between the terminals and hence the voltage is developed across it. This further drives the current (DC) to the DC loads, inverter, or battery charging circuit.

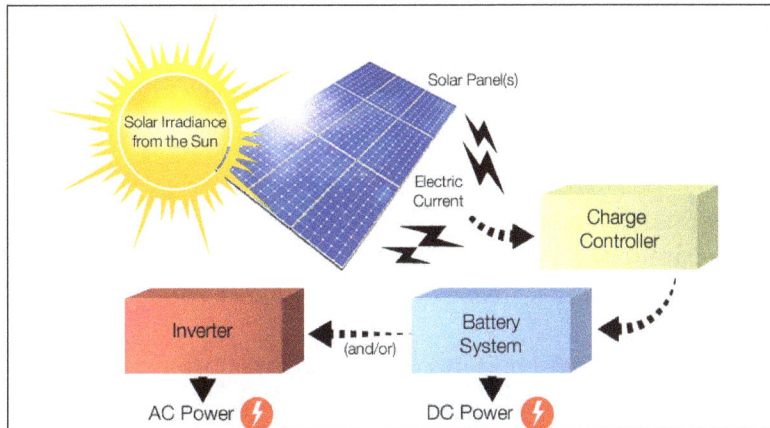

Photo electric effect

If more photons are absorbed, greater will be the current generated. However, much of the solar radiation fall on the solar cell is not converted into electricity.

This is because light is composed of photons of different wavelengths. Some photons hit the solar cell and then reflected and prevented from entering to the cell. In some materials, generated electrons recombine with other molecules before being drawn into current.

Likewise, there are many reasons for the low conversion rate or efficiency. The conversion efficiency of the solar panels used in individual residences ranges from 6 to 10%.

And for large-scale installations and solar power plants, solar panels are designed with best material and technologies to achieve higher conversion efficiency ranging from 40 to 60%, but they are costliest.

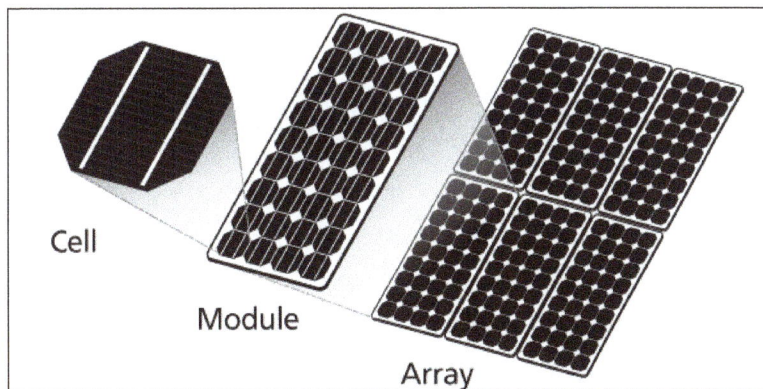

A single solar cell of 4 cm2 develops a voltage of 0.5 to 1 V and it can produce 0.7W power when exposed to the sunlight. Typically, the best designed solar panel has a maximum efficiency of 25%.

In order to generate high potential difference or voltage and more electric power, these individual cells are connected together that means some cells are connected in series and some are in parallel.

PV modules are formed by connecting the number of number of solar cells together. And several PV modules are connected together to make a PV array which can be used for small power as well as high power generation applications.

Components of Solar Energy Electricity Generation

The major components in solar energy electricity generation include:

Solar Panels

Solar cell modules or solar panels convert the solar energy into electricity. These are mounted in such a way that they collect maximum energy from the sun. Most solar panels are rated to a voltage 12V (a half volt PV cells are connected in series, inside of the solar panel to produce the high voltage say 12V).

Solar panels

The panels are connected in series to form a solar array that produces higher voltage, typically of 24 or 48V in standalone systems, or it can be several hundreds of volts in grid-connected systems.

If the panels are connected in parallel, the current delivered to the load will be more and hence the more power while maintaining the same voltage. Irrespective of series or parallel connection, the power rating of the system increases when multiple solar panels are connected together.

Here, a solar array is made with four 12V and 12 watt solar panels, where each panel produces a current of 1 A. Then this array would be rated as 48V, 48W with 1A current.

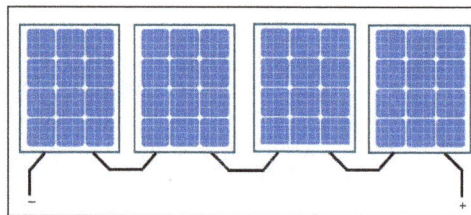

If the same rated solar panels are connected in parallel to form an array of four solar panels, then the solar array would be rated as 12V, 48W with 4 A current.

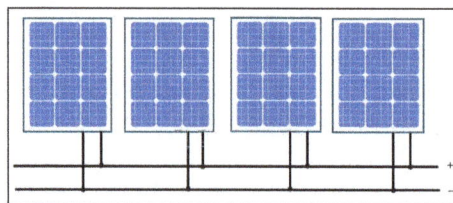

Batteries

Except the grid-connected system, all other solar energy power generation systems use batteries to store the energy generated from solar panels. Since the amount of solar power generated depends on the strength of the sunlight, batteries provide a constant source of power supply once it is fully charged.

Batteries

Mostly, lead acid batteries are used in solar electric systems. Like solar panels, batteries can be connected together to form a battery bank.

These can be connected either series or parallel as similar to solar panels to achieve desired voltage, current and power ratings. The type of battery chosen depends on the energy requirements of a system and its budget.

Controller

It regulates the flow of current into and out of the battery. If the generated current overcharges the battery, it leads to damage in the battery. Moreover, if the battery is completely discharged, it will destroy the battery. Hence the solar controller prevents the batteries to undergo these conditions.

Solar charge controller

The charge controller module balances the amount of electricity used to power appliance and lights with generated power. Also, it prevents the damage to the batteries due to overcharging and deep discharging. In addition, it gives the alarm when the module not functioning properly.

Inverter

Inverter

The electricity generated from the solar panels is a Direct Current (DC), whereas the most electrical appliances work on Alternating Current (AC) and hence a converter is needed to convert DC to AC, nothing but an inverter.

Also, if the solar system is connected to the grid, the generated DC voltage must be converted into AC. So the inverter equipment converts the DC voltage to the AC and to the same voltage as that of grid or appliance rating. As a recent invention, most individual solar panels are connected with micro inverters that provide a high AC voltage. These are suitable only for grid-connected systems and not suitable with battery backup systems.

Different Types of Solar Electric Systems

Off-grid/Standalone Solar Electric Systems

These are the most popular type of solar installations which are primarily designed to supplement or replace the conventional mains supply. These are mainly used in the locations where there is no other source to provide power supply and hence these are used in remote locations and rural areas where it is difficult to get the power from grid extensions.

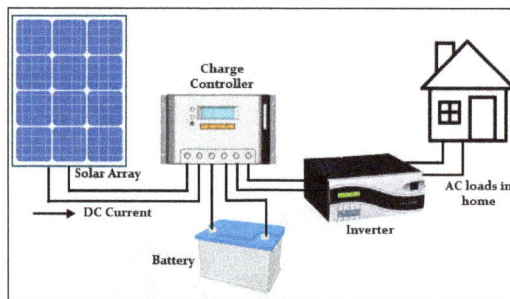

Generally, off-grid systems use solar power to charge the batteries, and this charge is then supplied to the load when needed. The battery power either directly operates the DC loads (DC lamps) or drives the power inverter that converts the DC power to AC power to operate the appliance like pumps, lighting equipment, refrigerators, etc.

This method is followed for any standalone system whether it is a pocket calculator or a complete off-grid home. Standalone systems are comparatively small and simple systems.

Grid-connected Solar Electric Systems

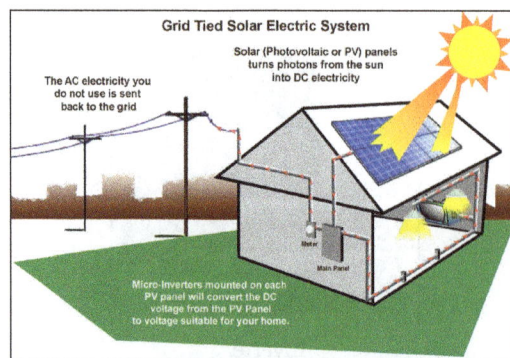

Grid-connected solar electric systems

These systems effectively create a micro power station and are connected directly into the electricity grid. These are normally found in urban areas where power is readily available.

During the day it feeds the excess electricity generated into the grid and during evening and night it imports the electricity from the grid. Here the grid acts like a storage medium in which power is taken from the grid when needed.

Grid-connected system doesn't have to supply enough electricity to cover entire power demand. So this system can be small or large depends on the owner's choice.

This system receives the payment for each kilowatt of power which is supplied to the electricity providers. This type of installation reduces the dependence on electric utilities and hence reduces the electricity bills.

The major components in this solar installation include a PV array or solar cells, inverter and the metering system. The main disadvantage of the grid-tied solar system is that it will switch off in the event of power cut, because this system is the part of utility grid and hence if there is a power cut, the power from the solar array is also switch off.

If this system won't stop, current flow back into the grid could lead to serious faults.

Central Grid-connected Solar Electric Systems

In a conventional grid system a variety of power sources such as gas, coal, water, etc. combined and it supplies the power to end user via transmission and distribution lines.

Likewise, a central grid connected system is directly connected to the transmission lines. These systems can be small (as 50kWp), large (as 100kWp) or somewhat higher range say 1GWp which all are directly connected to the transmission systems. Mostly, these are called as solar power plants.

Grid Fallback Solar Electric Systems

It combines the grid connected system with a bank of batteries. In this solar array generates the electricity, which in turn charges a battery bank. The battery power is then running the inverter which drives the loads through an inverter.

When the power in the batteries is not enough to drive the loads, the system automatically switches back to the grid power supply. Again, it switches back to battery power once the solar array recharges the batteries.

This system doesn't sell any power to any electrical utilities and the overall power generated used for domestic or residential systems alone.

Wind Power Generation

The wind is a source of free energy which has been used since ancient times in windmills for pumping water or grinding flour. The technology of high power, geared transmissions was developed centuries ago by windmill designers and the fantail wheel for keeping the main sales pointing into the wind was one of the world's first examples of an automatic control system.

Windmills at kinderdijk in the netherlands dating from 1740 used for pumping water from the polder

Though modern technology has made dramatic improvements to the efficiency of windmills which are now extensively use for electricity generation, they are still dependent on the vagaries of the weather. Not just on the wind direction but on the intermittent and unpredictable force of the wind. Too little wind and they can't deliver sufficient sustained power to overcome frictional losses in the system. Too much and they are susceptible to damage. Between these extremes, cost efficient installations have been developed to extract energy from the wind.

Available Power from the Wind

Theoretical Power

The power P available in the wind impinging on a wind driven generator is given by:

$P = \frac{1}{2}CA\rho v^3$

Where C is an efficiency factor known as the Power Coefficient which depends on the machine design, A is the area of the wind front intercepted by the rotor blades (the swept area), ρ is the density of the air (averaging 1.225 Kg/m³ at sea level) and v is the wind velocity.

The power is proportional to area swept by the blades, the density of the air and to the cube of the wind speed. Thus doubling the blade length will produce four times the power and doubling the wind speed will produce eight times the power.

Note also that the effective swept area of the blades is an annular ring, not a circle, because of the dead space around the hub of the blades.

Energy Conversion

Practical Power and Conversion Efficiency

German aerodynamicist Albert Betz showed that a maximum of only 59.3% of the theoretical power can be extracted from the wind, no matter how good the wind turbine is, otherwise the wind would stop when it hit the blades. He demonstrated mathematically that the optimum occurs when the rotor reduces the wind speed by one third.

In practical designs, inefficiencies in the design and frictional losses will reduce the power available from the wind still further. Converting this wind power into electrical power also incurs losses of up to 10% in the drive train and the generator and another 10% in the inverter and cabling. Furthermore, when the wind speed exceeds the rated wind speed, control systems limit the energy conversion in order to protect the electric generator so that ultimately, the wind turbine will convert only about 30% to 35% of the available wind energy into electrical energy.

Note that the power output from commercially available domestic wind turbines is usually specified at a steady, gust free, wind speed of 12.5 m/s. (Force 6 on the Beaufort scale corresponding to a strong breeze). In many locations, particularly urban installations, the prevailing wind will rarely reach this speed.

Blade Design for Optimum Energy Capture

Modern, high capacity wind turbines, such as those used by the electricity utilities in the electricity grid, typically have blades with a cross section similar to the aerofoils used to provide the lift in aircraft wings.

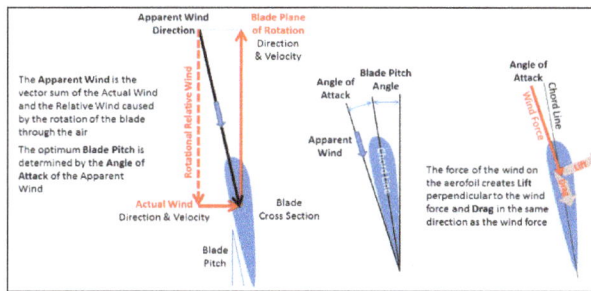

Turbine blade aerodynamics

The direction of the apparent wind that is the incident wind, relative to the chord line of the aerofoil is known as the angle of attack. Just as with aircraft wings, the lift resulting from the incident wind force increases as the angle of attack increases from 0 to a maximum of about 15 degrees at which point the smooth laminar flow of the air over the blade ceases and the air flow over the blade separates from the aerofoil and becomes turbulent. Above this point the lift force deteriorates rapidly while drag increases leading to a stall.

The tangential velocity S of any blade section at a distance r from the centre of rotation (the root of the blade) is given by $S = r\,\Omega$ where Ω is the angular velocity of rotation in radians.

For a given wind speed the apparent wind will be different at the root of the blade from the apparent wind at the tip of the blade because the rotational relative wind speed is different.

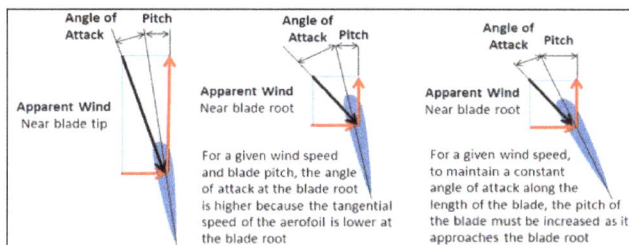

Angle of attack and blade twist

For a given speed of rotation, the tangential velocity of sections of the blade increases along the length of the blade towards the tip, so that the pitch of the blade must be twisted to maintain the same, optimum angle of attack at all sections along the length of the blade. The blade twist is thus optimised for a given wind speed. As the wind speed changes however, the twist will no longer be optimum. To retain the optimum angle of attack as wind speed increases a fixed pitch blade must increase its rotational speed accordingly, otherwise, for fixed speed rotors, variable pitch blades must be used.

The number of blades in the turbine rotor and its rotational speed must be optimised to extract the maximum energy from the available wind.

While using rotors with multiple blades should capture more wind energy, there is a practical limit to the number of blades which can be used because each blade of a spinning rotor leaves turbulence in its wake and this reduces the amount of energy which the following blade can extract from the wind. This same turbulence effect also limits the possible rotor speeds because a high speed rotor does not provide enough time for the air flow to settle after the passage of a blade before the next blade comes along.

There is also a lower limit to both the number of blades and the rotor speed. With too few rotor blades, or a slow turning rotor, most of the wind will pass undisturbed through the gap between the blades reducing the potential for capturing the wind energy. The fewer the number of blades, the faster the wind turbine rotor needs to turn to extract maximum power from the wind.

The notion of the Tip Speed Ratio (TSR) is a concept used by wind turbine designers to optimise a blade set to the shaft speed required by a particular electricity generator while extracting the maximum energy from the wind.

The tip speed ratio is given by:

$$TSR = \Omega R / V$$

Where Ω is the angular velocity of the rotor, R is the distance between the axis of rotation and the tip of the blade, and V is the wind speed.

A well designed typical three-bladed rotor would have a tip speed ratio of around 6 to 7.

Design Limits

For safety and efficiency reasons wind turbines are subject to operating limits depending on the wind conditions and the system design.

- Cut - in Wind Speed This is the minimum wind velocity below which no useful power output can be produced from wind turbine, typically between 3 and 4 m/s (10 and 14 km/h, 7 and 9 mph).

- Rated Wind Speed (also associated with the Nameplate Capacity): This is the lowest wind velocity at which the turbine develops its full power. This corresponds to the maximum, safe electrical generating capacity which the associated electrical generator can handle, in other words the generator's rated electrical power output. The rated wind speed is typically about 15 m/s (54 km/h, 34 mph) which is about double the expected average speed of the wind. To keep the turbine operating with wind speeds above the rated wind speed, control systems may be used to vary the pitch of the turbine blades, reducing the rotation speed of the rotor and thus limiting the mechanical power applied to the generator so that the electrical output remains constant. Though the turbine works with winds speeds right up to the cut-out wind speed, its efficiency is automatically reduced at speeds above the rated speed so that it captures less of the available wind energy in order to protect the generator. While it would be possible to use larger generators to extract full power from the wind at speeds over the rated wind speed, this would not normally be economical because of the lower frequency of occurrence of wind speeds above the rated wind speed.

- Cut - out Wind Speed: This is the maximum safe working wind speed and the speed at which the wind turbine is designed to be shut down by applying brakes to prevent damage to the system. In addition to electrical or mechanical brakes, the turbine may be slowed down by stalling or furling.

   o Stalling: This is a self-correcting or passive strategy which can be used with fixed speed wind turbines. As the wind speed increases so does the wind angle of attack until it reaches its stalling angle at which point the "lift" force turning the blade is destroyed. However increasing the angle of attack also increases the effective cross section of the blade face-on to the wind, and thus the direct wind force and the associated stress on the blades. A fully stalled turbine blade, when stopped, has the flat side of the blade facing directly into the wind.

   o Furling or Feathering: This is a technique derived from sailing in which the pitch control of the blades is used to decrease the angle of attack which in turn reduces the "lift" on the blades as well as the effective cross section of the aerofoil facing into the wind. A fully furled turbine blade, when stopped, has the edge of the blade facing into the wind reducing the wind force and stresses on the blade.

The cut-out speed is specified to be as high possible consistent with safety requirements and practicality in order to capture as much as possible of the available wind energy over the full spectrum of expected wind speeds. A cut-out speed of 25 m/s (90 km/h, 56 mph) is typical for very large turbines.

- Survival Wind Speed: This is the maximum wind speed that a given wind turbine is designed to withstand above which it cannot survive. The survival speed of commercial wind turbines is in the range of 50 m/s (180 km/h, 112 mph) to 72 m/s (259 km/h, 161 mph). The most common survival speed is 60 m/s (216 km/h, 134 mph). The safe survival speed depends on local wind a condition is usually regulated by national safety standards.

Yaw Control

Windmills can only extract the maximum power from the available wind when the plane of rotation of the blades is perpendicular to the direction of the wind. To ensure this the rotor mount must be free to rotate on its vertical axis and the installation must include some form of yaw control to turn the rotor into the wind.

For small, lightweight installations this is normally accomplished by adding a tail fin behind the rotor in line with its axis. Any lateral component of the wind will tend to push the side of the tail fin causing the rotor mount to turn until the fin is in line with the wind. When the rotor is facing into the wind there will be no lateral force on the fin and the rotor will remain in position. Friction and inertia will tend to hold it in position so that it does not follow small disturbances.

Large turbine installations have automatic control systems with wind sensors to monitor the direction of the wind and a powered mechanism to drive the rotor into its optimum position.

Capacity Factor

Electrical generating equipment is usually specified at its rated capacity. This is normally the maximum power or energy output which can be generated in optimal conditions. Since a wind turbine rarely works at its optimal capacity the actual energy output over a year will be much less than its rated capacity. Furthermore there will often be periods when the wind turbine cannot deliver any power at all. These occur when there is insufficient wind to power the turbine system, or other periods, fortunately only a few, when the wind turbine must be shut down because the wind speed is dangerously high and exceeds the system cut-out speed.

The capacity factor is simply the wind turbine generator's actual energy output for a given period divided by the theoretical energy output if the machine had operated at its rated power output for the same period. Typical capacity factors for wind turbines range from 0.25 to 0.30. Thus a wind turbine rated at 1 MegaWatt will deliver on average only about 250 kiloWatts of power.

Wind Supply Characteristics

Wind Speed

Though the force and power of the wind are difficult to quantify, various scales and descriptions have been used to characterise its intensity. The Beaufort scale is one measure in common use. The lowest point or zero on the Beaufort scale corresponds to the calmest conditions when the wind speed is zero and smoke rises vertically. The highest point is defined as force 12 when the wind speed is greater than 34 metres per second (122 km/h, 76 mph). As occurs in tropical cyclones when the countryside is devastated by hurricane conditions.

Small wind turbines generally operate between force 3 and force 7 on the Beaufort scale with the rated capacity commonly being defined at force 6 with a wind speed of 12 m/s.

Below force 3 the wind turbine will not generate significant power.

At force 3, wind speeds range from 3.6 to 5.8 m/s (8 to 13 mph). Wind conditions are described as "light" and leaves are in movement and flags begin to extend.

At force 7, wind speeds range from 14 to 17 m/s (32 to 39 mph). Wind conditions are described as "strong" and whole trees are in motion.

With winds above force 7 small, domestic wind turbines should be shut down to prevent damage.

Large turbines used in the electricity grid are designed to work with wind speeds of up to 25 m/s (90 km/h, 56 mph) which corresponds to between force 9 (severe gale, 23 m/s) and force 10 (storm, 27 m/s) on the Beaufort Scale.

Wind Consistency

Wind power has the advantage that it is normally available 24 hours per day, unlike solar power which is only available during daylight hours. Unfortunately the availability of wind energy is less predictable than solar energy. At least we know that the sun rises and sets every day. Nevertheless, based on data collected over many years, some predictions about the frequency of the wind at various speeds, if not the timing, are possible.

Wind Speed Distribution

Care should be taken in calculating the amount of energy available from the wind as it is quite common to overestimate its potential. You cannot simply take the *average* of the wind speeds throughout the year and use it to calculate the energy available from the wind because its speed is constantly changing and its power is proportional to the cube of the wind speed. (Energy = Power X Time). You have to weigh the probability of each wind speed with the corresponding amount of energy it carries.

Experience shows that for a given height above ground, the frequency at which the wind blows with any particular speed follows a Rayleigh Distribution. An example is shown in the figure.

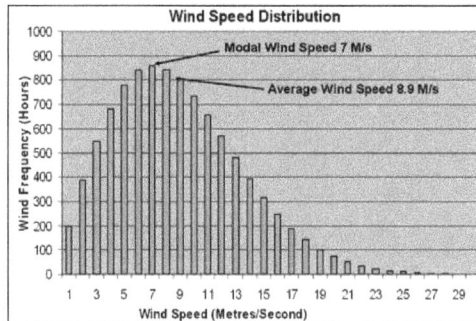

Wind Energy Distribution

The histogram below shows the resulting distribution of the wind energy content superimposed on the Rayleigh wind speed distribution (above) which caused it. Unfortunately not all of this wind energy can be captured by conventional wind turbines.

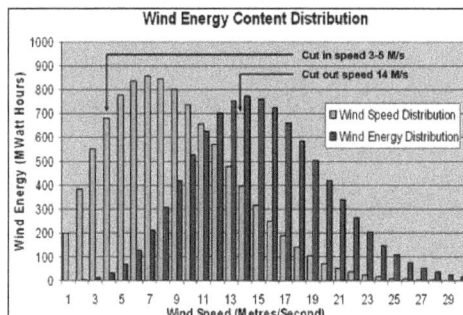

1. The peak wind energy occurs at wind speeds considerably above both the modal and average wind speeds since the wind energy content is proportional to the cube of its speed.

2. Very little energy is available at low speeds and most of this will be needed to overcome frictional losses in the wind turbine. Energy generation typically does not cut in until wind is blowing at speeds of at least 3 m/s to 5 m/s.

3. High wind speeds cause high rotation speeds and high stresses in the wind turbine which can can result in serious damage to the installation. To avoid these dangerous conditions, wind turbines are usually designed to cut out at wind speeds of around 25 m/s either by braking or feathering the rotor blades allowing the wind to spill over the blades, though smaller domestic installations may have lower operating limits.

4. Because of the limitations of the generating system and also upper speed limit at which the wind turbine can safely be used, it may capture only half or less of the available wind energy.

For a given wind speed the wind energy also depends on the elevation of the wind turbine above sea level. This is because the density of the air decreases with altitude and the wind energy is proportional to the air density. This effect is shown in the following histogram.

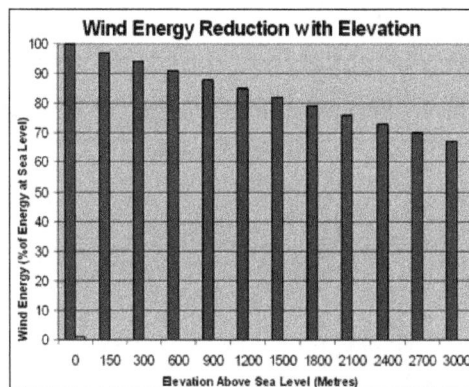

1. For a given wind speed the wind energy density decreases with increases in altitude. However at the same time the actual wind speeds tend to increase with height above ground level. Since the wind energy is proportional to the cube of the wind speed, the net effect is that wind energy tends to increase with the height above ground level.

2. As the density of air decreases with altitude, the wind energy density also decreases. By contrast the available solar energy increases with altitude due to lower atmospheric absorption.

Location Considerations

Generally marine locations and exposed hilltops provide the most favourable wind conditions with wind speeds consistently greater than 5 m/s.

Turbulent conditions will reduce the amount of energy which can be extracted from the wind reducing in turn the overall efficiency of the system. This is more likely to be the case over land than over the sea. Raising the height of the turbine above the ground effectively lifts it above the worst of the turbulence and improves efficiency.

Domestic wind turbines located between buildings in urban environments rarely operate at peak efficiency suffering from turbulence as well as being shielded from the wind by buildings and trees.

Practical Systems

Community/Grid Installations

Vesta 7 mw wind turbines with a rotor diameter of 164 m

Grid connected systems are dimensioned for average wind speeds 5.5 m/s on land and 6.5 m/s offshore where wind turbulence is less and wind speeds are higher. While offshore plants benefit from higher sustainable wind speeds, their construction and maintenance costs are higher.

Large scale wind turbine generators with outputs of up to 8 MWe or more with rotor diameters up to164 metres are now functioning in many regions of the world with even larger designs in the pipeline.

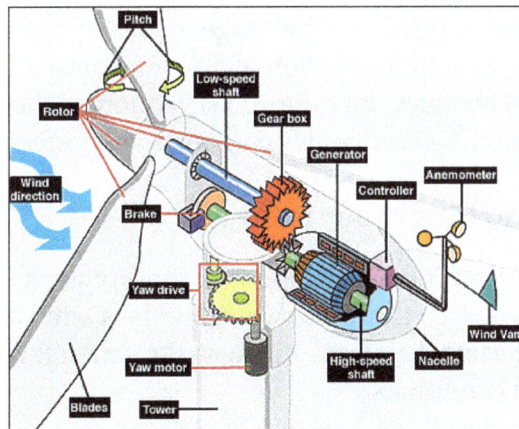

Large rotor blades are necessary to intercept the maximum air stream but these give rise to very high tip speeds. The tip speeds however must be limited, mainly because of unacceptable noise levels, resulting in very low rotation speeds which may be as low as 10 to 20 rpm for large wind turbines. The operating speed of the generator is however is much higher, typically 1200 rpm, determined by the number of its magnetic pole pairs and the frequency of the grid electrical supply. Consequently a gearbox must be used to increase the shaft speed to drive the generator at the fixed synchronous speed corresponding to the grid frequency.

Note that a "synchronous generator" is one whose electrical output frequency is synchronised to its shaft speed. It is not necessarily synchronised to the grid frequency, although that is usually an objective and extra, external controls are necessary to achieve this.

Fixed Speed Wind Turbine Generators

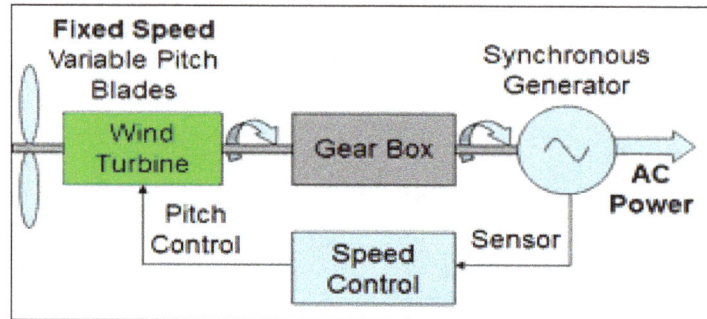

Large scale wind power

A typical fixed speed system employs a rotor with three variable pitch blades which are controlled automatically to maintain a fixed rotation speed for any wind speed. The rotor drives a synchronous generator through a gear box and the whole assembly is housed in a nacelle on top of a substantial tower with massive foundations requiring hundreds of cubic metres of reinforced concrete.

Fixed speed systems may however suffer excessive mechanical stresses. Because they are required to maintain a fixed speed regardless of the wind speed, there is no "give" in the mechanism to absorb gusty wind forces and this result in high torque, high stresses and excessive wear and tear on the gear box increasing maintenance costs and reducing service life. At the same time, the reaction time of these mechanical systems can be in the range of tens of milliseconds so that each time a burst of wind hits the turbine, a rapid fluctuation of electrical output power can be observed. Furthermore, variable speed wind turbines can capture 8-15% more of the wind's energy than constant speed machines. For these reasons, variable speed systems are preferred over fixed speed systems.

Variable Speed Wind Turbine Generators

A variable speed generator is better able to cope with stormy wind conditions because its rotor can speed up or slow down to absorb the forces when bursts of wind suddenly increase the torque on the system. The electronic control systems will keep the generator's output frequency constant during these fluctuating wind conditions.

Synchronous Generator with In-Line Frequency Control

Rather than controlling the turbine rotation speed to obtain a fixed frequency synchronised with the grid from a synchronous generator, the rotor and turbine can be run at a variable speed corresponding to the prevailing wind conditions. This will produce a varying frequency output from the generator synchronised with the drive shaft rotation speed. This output can then be rectified in the generator side of an AC-DC-AC converter and the converted back to AC in an inverter in grid side of the converter which is synchronised with the grid frequency. The grid side converter can also be used to provide reactive power (VArs) to the grid for power factor control and voltage regulation by varying the firing angle of the thyristor switching in the inverter and thus the phase of the output current with respect to the voltage.

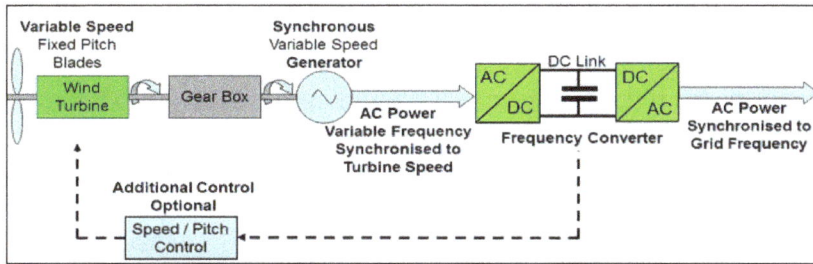

Large scale wind power in-line frequency conversion (grid systems)

The range of wind speeds over which the system can be operated can be extended and mechanical safety controls can be incorporated by means of an optional speed control system based on pitch control of the rotor vanes as used in the fixed speed system.

One major drawback of this system is that the components and the electronic control circuits in the frequency converter must be dimensioned to carry the full generator power. The doubly fed induction generator DFIG overcomes this difficulty.

Doubly Fed Induction Generator – DFIG

DFIG technology is currently the preferred wind power generating technology. The basic grid connected asynchronous induction generator gets its excitation current from the grid through the stator windings and has limited control over its output voltage and frequency. The doubly fed induction generator permits a second excitation current input, through slip rings to a wound rotor permitting greater control over the generator output.

The DFIG system consists of a 3 phase wound rotor generator with its stator windings fed from the grid and its rotor windings fed via a back to back converter system in a bidirectional feedback loop taking power either from the grid to the generator or from the generator to the grid.

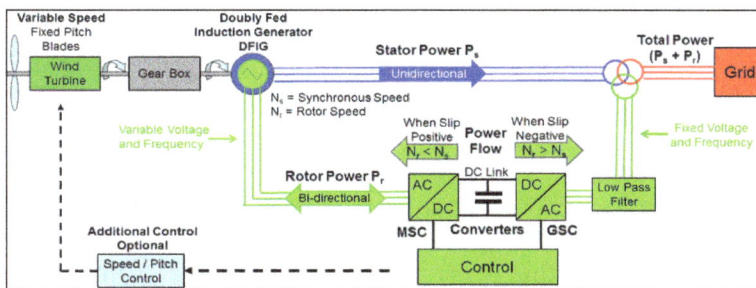

Asynchronous DFIG wind power generator (grid scale)

Generator Operating Principle

The feedback control system monitors the stator output voltage and frequency and provides error signals if these are different from the grid standards. The frequency error is equal to the generator slip frequency and is equivalent to the difference between the synchronous speed and the actual shaft speed of the machine.

The excitation from the stator windings causes the generator to act in much the same way as a basic squirrel cage or wound rotor generator. Without the additional rotor excitation, the frequency of a slow running generator will be less than the grid frequency which provides its excitation and its slip would be positive. Conversely if it was running too fast the frequency would be too high and its slip would be negative.

The rotor absorbs power from the grid to speed up and delivers power to the grid in order to slow down. When the machine is running synchronously the frequency of the combined stator and rotor excitation matches the grid frequency, there is no slip and the machine will be synchronised with the grid.

- Grid Side Converter - GSC: Carries current at the grid frequency. It is an AC to DC converter circuit used to provide a regulated DC voltage to the inverter in the machine side converter (MSC). It is used maintain a constant DC link voltage. A capacitor is connected across the DC link between the two converters and acts as an energy storage unit. The grid side converter is used to maintain a constant DC link voltage. In the opposite direction the GSC invereter delivers power to the grid with the grid regulated frequency and voltage.

  As with the in-line converter by adjusting the timing of the GSC inverter switching, the GSC converter also provides variable reactive power output to counterbalance the reactive power drawn from the grid enabling power factor correction as in the in-line frequency control system.

- Machine Side Converter - MSC: Carries current at slip frequency. It is an DC to AC inverter which is used to provide variable AC voltage and frequency to the rotor to control the torque and speed of the machine.

  When the generator is running too slowly, its frequency will be too low so that it is essentially motoring. The machine side converter takes DC power from the DC link and provides AC output power at the slip frequency to the rotor to eliminate its motoring slip and thus increase its speed. If the rotor is running too fast causing the generator frequency to be too high, the MSC extracts AC power from the rotor at the slip frequency causing it to slow down, reducing the generator slip, and converts the rotor output to DC passing it through the DC link to the GSC where it is converted to the fixed grid voltage and frequency and is inserted into the grid.

DFIG Control

1. Frequency: The frequency of the rotor currents induced by transformer action from the stator is the same as the slip frequency and this is equivalent to the frequency error signal in the feedback loop.

The additional direct excitation of the rotor adds a second set of controlled currents to the currents already induced in the rotor by transformer action from the stator. These additional currents affect the rotation speed of the rotor in the same way as the stator induced currents, producing an additional driving torque on the rotor except that the additional rotor currents are independent of the speed of the rotor. The frequency of the control current supplied by the MSC can be precisely controlled to match and thus neutralise the slip frequency so that, with zero slip, the generator rotates at the synchronous frequency determined by the grid. The greater the slip, the greater the compensating frequency required.

The control system has to respond to both positive (motor) slip and negative (generator) slip.

To increase the speed of a slow running rotor, the phase sequence of the rotor windings is set so that the rotor magnetic field is in the same direction as the generator rotor producing negative slip to counteract and thus neutralise the rotor's positive slip. To reduce the rotor speed, the phase sequence of the rotor windings is set in opposite direction from the generator's rotation producing positive slip to counteract the rotor's negative slip.

When operating at synchronous speed the rotor current will be DC current and there will be no sip and no power flow through the rotor.

2. Voltage: The generator output voltage is determined by the magnitude of the excitation current supplied to the rotor and this can be adjusted by means of the rotor input voltage provided by the MSC. A chopper or pulse width modulator PWM is used to generate the variable DC control voltage necessary. The converter feedback controls thus enable the excitation current to be regulated by the MSC to neutralise the voltage error signal and thus obtain a constant bus voltage matched to the grid voltage.

DFIG Performance

1. The DFIG system provides regulated power tied to the grid frequency and voltage when driven by varying levels of torque from the wind.

2. Typical speed control range is ± 30% of synchronous speed.

3. For a greater speed control range it may be necessary to implement separate pitch control on the wind turbine's rotor vanes.

4. The generator power flow is shared by the stator and the rotor with 70% or more coming from the stator. The feedback loop only carries the slip power which is between 20% and 30% of the total.

5. Because of the reduced power flowing through the converters, compared with the in-line control system above, they the DFIG converters can be implemented with less expensive lower power components.

6. The DFIG machine can produce up to twice the power of a similar sized singly fed machine while incurring similar losses; however the losses in the electronic controls must be added to this. Nevertheless the DFIG machine efficiency is better that a singly fed machine.

Hydroelectric Power Generation

Hydro energy is available in many forms, potential energy from high heads of water retained in dams, kinetic energy from current flow in rivers and tidal barrages, and kinetic energy also from the movement of waves on relatively static water masses. Many ingenious ways have been developed for harnessing this energy but most involve directing the water flow through a turbine to generate electricity. Those that don't usually involve using the movement of the water to drive some other form of hydraulic or pneumatic mechanism to perform the same task.

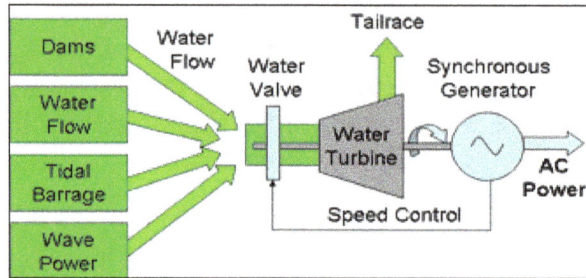

Hydro electric power generation

Water Turbines

Like steam turbines, water turbines may depend on the impulse of the working fluid on the turbine blades or the reaction between the working fluid and the blades to turn the turbine shaft which in turn drives the generator. Several different families of turbines have been developed to optimise performance for particular water supply conditions.

Turbine Power Output

In general, the turbine converts the kinetic energy of the working fluid, in this case water, into rotational motion of the turbine shaft. Swiss mathematician Leonhard Euler showed in 1754 that the torque on the shaft is equal to the change in angular momentum of the water flow as it is deflected by the turbine blades and the power generated is equal to the torque on the shaft multiplied by the rotational speed of the shaft.

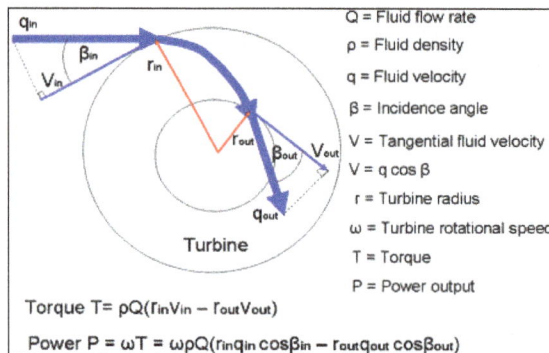

Q = Fluid flow rate
ρ = Fluid density
q = Fluid velocity
β = Incidence angle
V = Tangential fluid velocity
$V = q \cos \beta$
r = Turbine radius
ω = Turbine rotational speed
T = Torque
P = Power output

Torque $T = \rho Q(r_{in} V_{in} - r_{out} V_{out})$

Power $P = \omega T = \omega \rho Q(r_{in} q_{in} \cos \beta_{in} - r_{out} q_{out} \cos \beta_{out})$

Euler's turbine equation

Note that this result does not depend on the turbine configuration or what happens inside the turbine. All that matters is the change in angular momentum of the fluid between the turbine's input and output.

Hydroelectric Power Generation Efficiency

Hydroelectric power generation is by far the most efficient method of large scale electric power generation. Energy flows are concentrated and can be controlled. The conversion process captures kinetic energy and converts it directly into electric energy. There are no inefficient intermediate thermodynamic or chemical processes and no heat losses. The overall efficiency can never be 100% however since extracting 100% of the flowing water's kinetic energy means the flow would have to stop.

The conversion efficiency of a hydroelectric power plant depends mainly on the type of water turbine employed and can be as high as 95% for large installations. Smaller plants with output powers less than 5 MW may have efficiencies between 80 and 85 %. It is however difficult to extract power from low flow rates.

Turbine Types

The most appropriate turbine to use depends on the rate of water flow and the head or pressure of water.

Impulse Turbines

Impulse turbines require tangential water flow on one side of the turbine runner (rotor) and must therefore operate when only partly submerged. They are best suited to applications with a high head but a low volume flow rate such as fast flowing shallow water courses though it is used in a wide range of situations with heads from as low as 15 metres up to almost 2000 metres.

Pelton Turbine: The Pelton turbine is an example of an impulse turbine.High pressure heads give rise to very fast water jets impinging in the blades resulting in very high rotational speeds of the turbine. The split bucket pairs divide the water flow ensuring balanced axial forces on the turbine runner.

Pelton turbine

Pelton wheels are ideal for low power installations with outputs of 10kW or less but they have also been used in installations with power outputs of up to 200 MW. Efficiencies up to 95% are possible.

Reaction Turbines

Reaction turbines are designed to operate with the turbine runner fully submerged or enclosed in a casing to contain the water pressure. They are suitable for lower heads of water of 500 metres or less and they are the most commonly used high power turbines.

1. Francis Turbine: The Francis turbine is an example of a reaction turbine. Water flow enters in a radial direction towards the axis and exits in the direction of the axis.

Francis turbine

Large scale turbines used in dams are capable of delivering over 500 MW of power from a head of water of around 100 metres with efficiencies of up to 95%.

2. Propeller and Kaplan Turbines: The propeller turbine is another example of a reaction turbine. Designed to work fully submerged, it is similar in form to a ship's propeller and is the most suitable design for low head water sources with a high flow rate such as those in slow running rivers. Designs are optimised for a particular flow rate and efficiencies drop of rapidly if the flow rate falls below the design rating. The Kaplan version has variable pitch vanes to enable it to work efficiently over a range of flow rates.

Kaplan – variable pitch propeller turbine

Power from Dams (Potential Energy)

Supply Characteristics

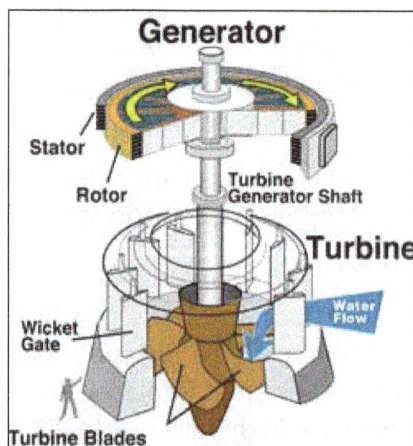

A hydroelectric dam installation uses the potential energy of the water retained in the dam to drive a water turbine which in turn drives an electric generator. The available energy therefore depends on the head of the water above the turbine and the volume of water flowing through it. Turbines are usually reaction types whose blades are fully submerged in the water flow.

The diagram opposite shows a typical turbine and generator configuration as used in a dam.

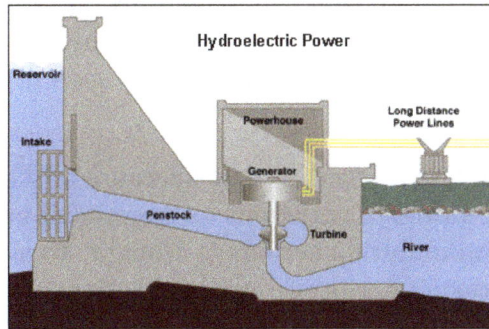

The civil works involved in providing hydro-power from a dam will usually be many times the cost of the turbines and the associated electricity generating equipment. Dams however provide a large water reservoir from which the flow of water, and hence the power output of the generator, can be controlled. The reservoir also serves as a supply buffer storing excess water during rainy periods and releasing it during dry spells.

The build-up of silt behind the dam can cause maintenance problems.

Available Power

Potential energy per unit volume = ρgh.

Where, ρ is he density of the water (10^3 Kg/m³), h is the head of water and g is the gravitational constant (10 m/sec²).

The power P from a dam is given by:

$$P = \eta\rho ghQ$$

Where Q is the volume of water flowing per second (the flow rate in m³/second) and η is the efficiency of the turbine.

For water flowing at one cubic metre per second from a head of one metre, the power generated is equivalent to 10 kW assuming an energy conversion efficiency of 100% or just over 9 kW with a turbine efficiency of between 90% and 95%.

Run of River Power (Kinetic Energy)

Supply Characteristics

"Run-of-river" installations do not depend on flooding large tracts of land to form dams. Instead, the necessary constant water supply may be derived from natural upstream lakes and reservoirs. They are typically used for smaller schemes generating less than 10 MegaWatts output power.

Water from a fast flowing river or stream is diverted through a turbine, often a Pelton wheel which drives the electrical generator. The local head of water may be essentially not much more than zero and the turbine is designed to convert the kinetic energy of the flowing water into the rotational energy of the turbine and the generator. The available energy therefore depends on the quantity of water flowing through the turbine and the square of its velocity.

Impulse turbines which are only partially submerged are more commonly employed in fast flowing run-of-river installations while in deeper, slower flowing rivers with a greater head of water, fully submerged Kaplan reaction turbines may be used to extract the energy from the water flow.

Run-of-river projects are much less costly than dams because of the simpler civil works requirements. They are however susceptible to variations in the rainfall or water flow which reduce or even cut off potential power output during periods of drought. To avoid the problems of seasonal river flows, or even daily fluctuations, run-of-river installations may incorporate an additional, limited amount of "man-made" water storage, referred to as "pondage", to keep the plant operating during dry periods.

On the other hand, during flood conditions the installation may not be able to accommodate the higher flow rates and water must be diverted around the turbine losing the potential generating capacity of the increased water flow.

Because of these limitations, if the construction of a dam is not possible, run of river installations may also need to incorporate some form of supply back-up such as battery storage, emergency generators or even a grid connection.

Available Power

The maximum power output from a turbine used in a run of river application is equal to the kinetic energy ($\frac{1}{2}mv^2$) of the water impinging on the blades. Taking the efficiency η of the turbine and its installation into account, the maximum output power P_{max} is given by:

$$P_{max} = \frac{1}{2}\eta\rho Qv^2$$

Where, v is the velocity of the water flow and Q is the volume of water flowing through the turbine per second.

Q is given by:

$$Q = Av$$

Where, A is the swept area of the turbine blades.

Thus,

$$P_{max} = \frac{1}{2}\eta\rho Av^3$$

This relationship also applies to shrouded turbines used to capture the energy of tidal flows and is directly analogous to the equation for the theoretical power generated by wind turbines. Note that the power output is proportional to the cube of the velocity of the water.

Thus the power generated by one cubic metre of water flowing at one metre per second through a turbine with 100% efficiency will be 0.5 kW or slightly less taking into account the inefficacies in the system. This is only one twentieth of the power generated by the same volume flow from the dam above. To generate the same power with the same volume of water from a run of river installation the speed of the water flow should be √20 metres per second (4.5 m/sec).

Tidal Power

Supply Characteristics

Harnessing the power of the tides can be achieved by placing bi-directional turbines in the path of the tidal water flow in bays and river estuaries. To be viable, it needs a large tidal range and involves creating a barrier across the bay or estuary to funnel the water through the turbines as the tide comes in and goes out.

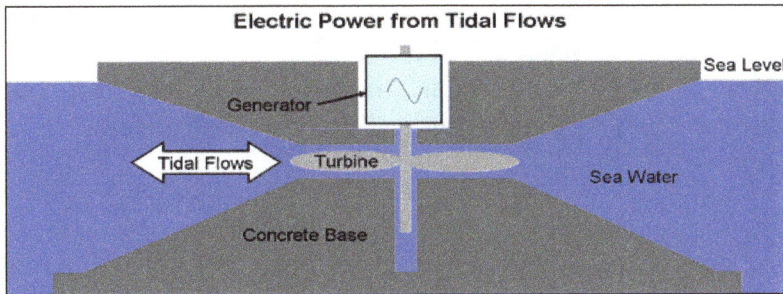

Electric power from tidal flows

Tidal power comes closest of all the intermittent renewable sources to being able to provide an unlimited, continuous and predictable power output but unfortunately there are few suitable sites in the world and environmental constraints have so far prevented their general acceptance.

Shrouded water turbines placed in deep water tidal currents show better potential for exploitation, though the associated civil works are more complicated, and several projects are under development.

Power is available for only six to twelve hours per day depending on the ebb and flow of the tides.

Available Power

The maximum power output from a shrouded water turbine used in tidal energy applications is equal to the kinetic energy of the water impinging on the blades, similar to the "run of river" calculation above. Taking the efficiency η of the turbine and its installation into account, the maximum output power P_{max} is given by:

$$P_{max} = \tfrac{1}{2}\eta\rho Av^3$$

Where, v is the velocity of the water flow and A is the swept area of the blades.

A turbine one metre in diameter with a water current of one metre per second flowing through it would generate 0.4 kW of electricity assuming 100% efficiency. Similarly a 3 meter diameter turbine with a water current of 3 metres per second would produce 32 Kw of power.

Wave Power

Supply Characteristics

The energy available from the ocean's surface wave motion is almost in limited, but it has proved frustratingly difficult to capture. Many ingenious systems have been proposed but, except for very small installations, very few are generating electricity commercially and most have been thwarted by practical problems.

Some of these proposals are outlined below. Most are still in an experimental phase and many are not scalable into high capacity systems.

Energy Conversion Systems

- Oscillating Float System: One of the simplest and most common solutions is the oscillating float system in which a float is housed inside an cylinder shaped buoy which is open at the bottom and moored to the seabed. Inside the cylinder the float moves up and down on the surface of the waves as they pass through the buoy. Various methods have been employed to turn the motion of the float into electrical energy. These include:

  o Hydraulic systems in which air is compressed in a pneumatic reservoir above the float during its upward movement on the crests of the waves. After the crests have passed, the air expands and forces the float downwards into the following troughs of the waves. A hydraulic system then uses the reciprocating movement of the float to pump water through a water turbine which drives a rotary electrical generator.

  o Pneumatic systems in which the air displaced in the cylinder is used to power an air turbine which drives the generator.

  o Linear generators to turn the reciprocating motion of the float directly into electrical power.

  o Instead of generating the electricity on board the buoy, some systems pump the hydraulic fluid ashore to power shore based generators.

- Oscillating Paddle System: This system uses large paddles moored to the ocean floor to mimic the swaying motion of sea plants in the presence of ocean waves. The paddles are fixed to special hinged joints at the base which use the swaying motion of the paddles to pump water through a turbine generator.

- Oscillating Snake System: The snake system uses a series of floating cylindrical sections linked by hinged joints. The floating snake is tethered to the sea bed and maintains a position head on into the waves. The wave-induced motion at the hinges is used to pump high-pressure oil through hydraulic motors via smoothing accumulators. The hydraulic motors in turn drive electrical generators to produce the electrical power.

- Oscillating Water Column: Water columns are formed within large concrete structures built on the shore line or on rafts. The structure is open at both the top and the bottom. The lower end is submerged in the sea and an air turbine fills the aperture at the top. The rising and falling of the water column inside the structure moves the air column above it driving the air through the turbine generator. The turbine has movable vanes which rotate to maintain unidirectional rotation when the movement of the air column reverses.

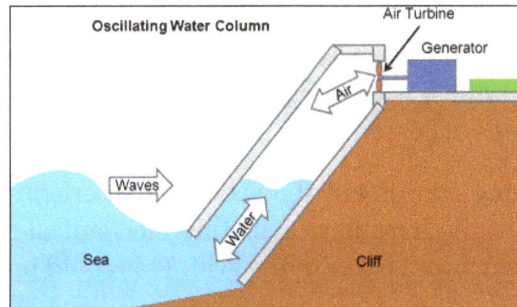

- Pressure Transducer System: The hydraulic pump system uses a submerged gas-filled tank with rigid sides and base and a flexible, bellows-like, top. The gas in the tank compresses and expands in response to pressure changes from the waves passing overhead causing the top to rise and fall. A lever attached to centre of the top drives pistons, which pump pressurized water ashore for driving hydraulic generators.

- Wave Capture Systems: Wave capture systems use a narrowing ramp to funnel waves into an elevated reservoir. Waves entering the funnel over a wide front are concentrated into a narrowing channel which causes the amplitude of the wave to increase. The increased wave height coupled with the momentum of the water is sufficient to raise a quantity of water up a ramp and into a reservoir situated above the sea level. Water form the reservoir can then be released through a hydroelectric turbine located below the reservoir to generate electricity.

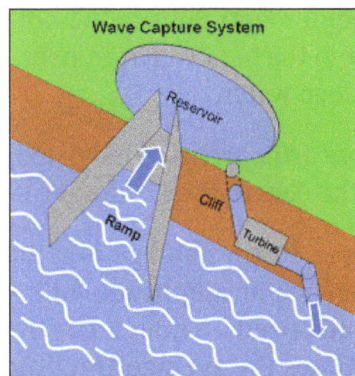

- Overtopping Wave Systems: These are floating systems similar to the land based system. They focus waves onto a tapered ramp which causes their amplitude to increase. The crests of the waves overtop the ramp and spill into a low dam. Water from the low dam then flows through hydroelectric turbines back into into the sea beneath the floating structure.

- Lever Systems: Various lever based energy capture systems have been developed. Long levers may be mounted on steel piles or on floating platforms. Large floats or buoys are attached to the extremities of the levers which move up and down with the waves. Movement of the lever arms forces fluid into a central hydraulic accumulator and through to a generator turbine. Alternatively high-pressure water can be pumped ashore to power shore based generators.

Technical Challenges

Formidable technical challenges are involved in designing practical systems for capturing wave energy.

- Variability of the sea conditions: Sea conditions are notoriously variable and the system must be able to cope with a wide range of wave amplitudes and frequencies as well as changes in the directions of currents.

- Matching the generating equipment the wave characteristics: Mechanisms are required to convert the power of the irregular oscillating mechanical forces induced by the waves into electrical power synchronised with the grid. This could involve some expensive power electronics.

  Typical rotating machines used for power generation operate at a synchronous speed of 1200 r.p.m. (20 revolutions per second) whereas the frequency of waves driving the generator is likely to be between 5 and 10 seconds per cycle. A mechanical gearing system is needed to match this 200:1 ratio in operating speeds, possibly combined with special purpose, slow speed generators, incorporating a large number of pole pairs.

  One way around all of these problems is to use hydraulic accumulators either in situ or on shore to smooth out the energy delivery to the generator.

- Equipment construction: For reasonable sized systems, very high mechanical forces will be involved converting the wave energy into mechanical energy for driving the electrical generator.

- Housing and mooring the equipment: Substantial housings must be provided to protect the generating equipment from the harsh environment. Holding the installation in place is also particularly difficult in deep water.

- Energy transmission: Low loss armoured and insulated cables or high pressure pipes must be developed for delivering the electrical or hydraulic energy back to the shore.

- Resistance to storm damage: Storm damage is a major threat. The frequency of occurrence of waves of any particular amplitude follows a Rayleigh distribution similar to that which applies to wind speeds. Though the frequency of serious storms may be rather small, a wave of ten times the average amplitude may be expected once every 50 years. From the power calculation below, the wave power is proportional to the square of the wave amplitude. This means that the installation must be designed to withstand forces one hundred times greater than the normal working level. This adds considerably to the costs.

Available Power

The wave power per unit length of the wave front P_L is given by (Twiddel & Weir. Renewable Energy Resources) as:

$$P_L = \rho g a^2 \lambda / 4T$$

Where ρ is he density of the water (10^3 Kg/m^3), a is the wave amplitude (half of the wave height), g is the gravitational constant (10 m/sec^2), λ is the wave length of the oscillation and T the period of the wave.

Thus for a wave with amplitude 1.5 metres, length 100 metres and period 5 seconds, the power per metre of wave front will be 75 kW.

Geothermal Power Generation

Geothermal power plants use steam produced from reservoirs of hot water found a few miles or more below the Earth's surface to produce electricity. The steam rotates a turbine that activates a generator, which produces electricity. There are three types of geothermal power plants: dry steam, flash steam, and binary cycle.

Dry Steam Plants

One of 22 dry steam plants at the geysers in california

These plants use dry steam that is naturally produced in the ground. This steam travels from the production well to the surface and through a turbine, and after transferring its energy to the turbine it condenses and is injected back into the Earth. These types are the oldest types of geothermal power plants, because this type of power plant requires the highest temperatures they can only be used where the temperature underground is quite high, but this type requires the least fluid flow.

Flash Cycle Steam Plants

These types are the most common due to the lack of naturally occurring high-quality steam. In this method, water must be over 180°C, and under its own pressure it flows upwards through the well. This is a lower temperature than dry steam plants have. As its pressure decreases, some of the water "flashes" to steam, which is passed through the turbine section. The remaining water that did not become steam is cycled back down into the well, and can also be used for heating purposes. The cost of these systems is increased due to more complex parts, however they can still compete with conventional power sources.

Binary Cycle Plants

Binary power plants are expected to be the most commonly used type of geothermal power plant in the future, as locations outside of the known hot spots begin to use geothermal energy. This is because binary cycle plants can make use of lower temperature water than the other two types of plants. They use a secondary loop (hence the name "binary") which contains a fluid with a low boiling point, such as pentane or butane. The water from the well flows through a heat exchanger which transfers its heat to this fluid, which vaporizes due to its low boiling point. It is then passed through a turbine, accomplishing the same task as steam.

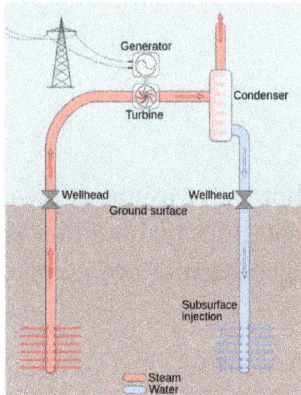

Dry steam cycle

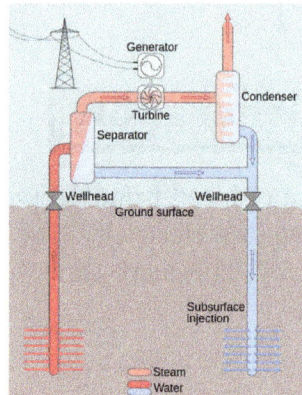

Flash steam cycle

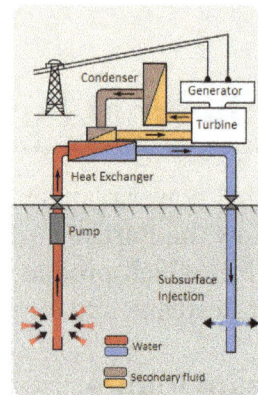

Binary cycle: The lighter brown is vaporized butane, while darker brown is liquid butane

Steam Turbine Power Generation

The steam turbine is one kind of heat engine machine in which steam's heat energy is converted to mechanical work. The construction of steam turbine is very simple. There is no piston rod, flywheel or slide valves attached to the turbine. So maintenance is quite easy. It consists of a rotor and a set of rotating blades which are attached to a shaft and the shaft is placed in the middle of the rotor. An electric generator known as steam turbine generator is connected to the rotor shaft. The turbine generator collects the mechanical energy from the shaft and converts it into electrical energy. Steam turbine generator also improves the turbine efficiency.

Types of Steam Turbine

According to the working principle, there are different types of steam turbine:

1. According to the working principle steam turbines are mainly divided into two categories:

   o Impulse Turbine,

   o Reaction Steam Turbine.

When steam strikes the moving blades through nozzles called Impulse Turbine and when it strikes the moving blades under pressure via guide mechanism called Reaction Turbine.

2. According to the direction of steam flow, it may be classified into two categories:

   o Axial Flow Steam Turbine,

o   Radial Flow Steam Turbine.

When the flow of steam inside the casing is parallel to the rotor shaft axis then it is called Axial Flow Steam Turbine and flow of steam inside the casing is radial to the rotor shaft axis called Radial Flow Steam turbine.

3. According to the exhaust condition of steam, it is further divided into two categories:

o   Back Pressure or Non-Condensing types Steam Turbine,

o   Condensing type Steam Turbine.

After expansion of steam it is exhausted into atmosphere called back pressure steam turbine or non-condensing types steam turbine otherwise it exhausted into a condenser called condensing turbine.

4. According to pressure of steam, it may be divided following categories:

o   High-pressure or pass-out or Extraction steam turbine,

o   Medium-pressure or back pressure steam turbine,

o   Low-pressure turbine.

High, medium and low-pressure steam is supplied into the turbine, called high-pressure steam turbine or medium pressure steam turbine or back pressure steam turbine and low- pressure steam turbine.These turbines are used for various manufacturing and heating process.

5. According to the number of stages, it may be divided following categories:

o   Single stage steam turbine,

o   Multi-stage steam turbine.

Steam is coming from nozzles when passed through a single set of moving blades called single stage steam turbine and to flow multi-stages of moving blades called multi-stages steam turbine.

6. According to the blade and wheels arrangement, it may be divided following categories:

o   Pressure Compounding Steam Turbine,

o   Velocity Compounding Steam Turbine,

o   Impulse-Reaction Combined Steam Turbine,

o   Pressure-Velocity Compounding Steam Turbine.

Electrical energy generation using steam turbines involves three energy conversions, extracting thermal energy from the fuel and using it to raise steam, converting the thermal energy of the steam into kinetic energy in the turbine and using a rotary generator to convert the turbine's mechanical energy into electrical energy.

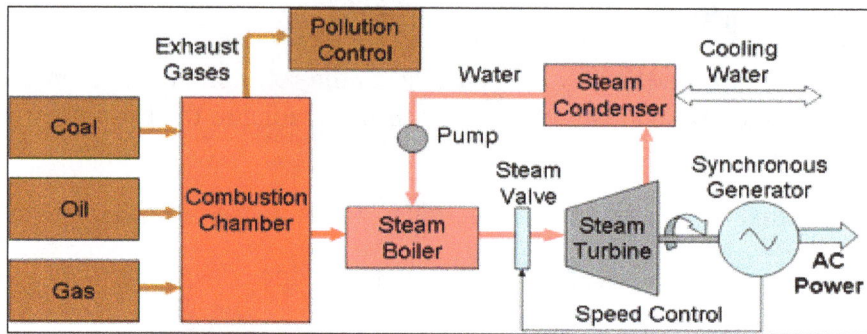

Fossil fuel powered steam turbine electricity generation

Raising Steam (Thermal Sources)

Steam is mostly raised from fossil fuel sources, three of which are shown in the above diagram but any convenient source of heat can be used.

1. Chemical Transformation: In fossil fuelled plants steam is raised by burning fuel, mostly coal but also oil and gas, in a combustion chamber. Recently these fuels have been supplemented by limited amounts of renewable biofuels and agricultural waste.

   The chemical process of burning the fuel releases heat by the chemical transformation (oxidation) of the fuel. This can never be perfect. There will be losses due to impurities in the fuel, incomplete combustion and heat and pressure losses in the combustion chamber and boiler. Typically these losses would amount to about 10% of the available energy in the fuel.

2. Nuclear Power: Steam for driving the turbine can also be raised by capturing the heat generated by controlled nuclear fission.

3. Solar Power: Similarly solar thermal energy can be used to raise steam, though this is less common.

4. Geothermal Energy: Steam emissions from naturally occurring aquifers are also used to power steam turbine power plants.

Steam Turbine Working Principles

High pressure steam is fed through a set of fixed nozzles in the turbine stator to the turbine rotor (runner) and passes along the machine axis through multiple rows of alternately fixed and moving blades. From the steam inlet port of the turbine towards the exhaust point, the blades and the turbine cavity are progressively larger to allow for the expansion of the steam.

The stator blades in each stage act as nozzles in which the steam expands and emerges at an increased speed but lower pressure. As the high velocity steam impacts on the moving blades it imparts some of its kinetic energy to the moving blades.

There are two basic steam turbine types, impulse turbines and reaction turbines, whose blades are designed control the speed, direction and pressure of the steam as is passes through the turbine.

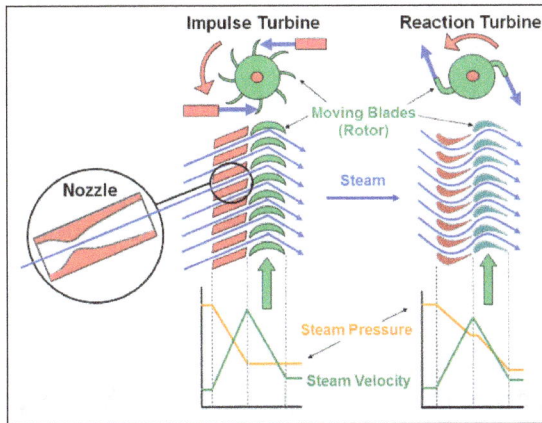

1. Nozzles: Key to achieving high efficiency in both impulse and reaction turbines is the design of the nozzles. They are normally a convergent-divergent (hourglass) shape which increases the velocity of the inlet steam while reducing its pressure. Increasing the velocity of the steam by means of a flared nozzle output orifice may seem counter-intuitive since water flows faster through a constricted part of a stream or a pipe and squeezing the end of a hosepipe causes the water to squirt out in a long, fast jet. This occurs because the water is an incompressible fluid. Steam on the other hand is a gas and its volume is not fixed but depends on its temperature and pressure. Gas dynamics are therefore quite different from hydrodynamics, however the conservation of energy principle still holds for both fluids and Bernoulli's Law indicates that kinetic energy of a gas increases as pressure energy falls.

This flared nozzle design was discovered by de Laval and applies equally to the nozzles of rocket engines whose working fluid is hot exhaust gas.

2. Impulse Turbines: The steam jets in an impulse turbine are directed by the fixed nozzles at the turbine's bucket shaped rotor blades where the force exerted by the jets causes the rotor to turn while at the same time the velocity of the steam is reduced as it imparts its kinetic energy to the blades. The blades in turn change change the direction of flow of the steam and this change of momentum corresponds to the increased momentum of the rotor. (Descartes - Conservation of momentum). The entire pressure drop in the turbine stage occurs in the fixed nozzles in the stator and there is no pressure drop as the steam passes through the rotor blades since the cross section of the chamber between the blades is constant. Impulse turbines are therefore also known as constant pressure turbines.

Steam impulse turbines usually operate at extremely high speeds of 30,000 r.p.m. or more and are thus subject to enormous centrifugal forces. For most practical applications the speed must be geared down. Other than that, the design is relatively simple and the turbine casing does not necessarily need to be pressure proof.

In a compound turbine, the next series of fixed blades reverses the direction of the steam before it passes to the second row of rotor blades.

3. Reaction Turbines: Both the fixed and the rotor blades of the reaction turbine are shaped more like aerofoils, arranged such that the cross section of the blades diminishes from the inlet side towards the exhaust side of the blades. This means that the cross section of the steam passages

between both sets of fixed and rotor blades increases across the turbine stage. In this way both sets of blades essentially form nozzles so that as the steam progresses through both the stator and the rotor its pressure decreases causing its velocity to increase. The rotor becomes basically a set of rotating nozzles.

As the steam emerges in a jet from between each set of rotor blades, it creates a reactive force on the blades which in turn creates the turning moment on the turbine rotor, just as in Hero's steam engine. (Newton's Third Law - For every action there is an equal and opposite reaction).

Reaction turbines are generally much more efficient than impulse turbines and run at lower speeds which mean they don't necessarily need reduction gearing. They are however more complex and the high pressure steam make them more susceptible to leaks between the stages.

4. Compound Steam Turbines: The compound turbine uses a series of turbine stages in which the output steam from each stage feeds into the next stage. By suitably shaping the rotor and stator blades to form nozzles the steam pressure or velocity can be reduced gradually over the series of stages rather than in a single stage. This permits the use of very high steam pressures and velocities enabling very high turbine power outputs.

Pressure Compounding

Pressure compounding uses a series of reaction turbine stages to solve the problem of very high blade velocity in the single-stage impulse turbines. The steam pressure drops across each stage as it gives up its pressure energy while the steam velocity remains fairly constant, changing direction as it passes through each stage. Because the steam pressure drops with each stage of the turbine, the volume of the steam increases correspondingly with every stage so that in high capacity turbines the blades and the turbine casing must in turn be correspondingly bigger for each subsequent lower pressure stage to accommodate this higher volume flow.

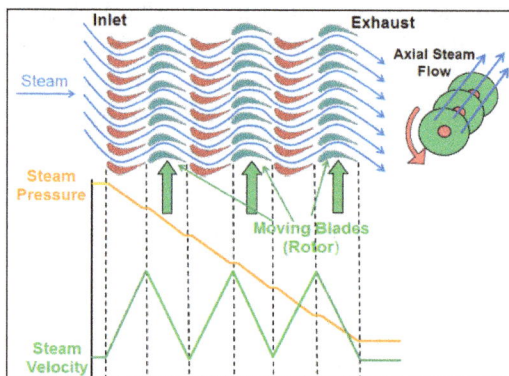

Three stage pressure compounded steam turbine

Impulse turbines are also compounded in a similar way however most turbines use a combination of impulse and reaction stages.

Velocity Compounding

Velocity compounding uses a series of impulse turbine stages. The input nozzles direct high velocity steam onto the first set of moving blades and as steam flows over the blade it imparts some of its momentum to blades losing some velocity, giving up its kinetic energy to the moving blades. There is no change in velocity of steam as it passes through the fixed blades. In this way the velocity of the steam reduces as it passes through the sets of moving blades of the turbine while the steam pressure remains fairly constant across the turbine.

The Condenser

The exhaust steam from the low pressure turbine is condensed to water in the condenser which extracts the latent heat of vaporization from the steam. This causes the volume of the steam to go to zero, reducing the pressure dramatically to near vacuum conditions thus increasing the pressure drop across the turbine enabling the maximum amount of energy to be extracted from the steam. The condensate is then pumped back into the boiler as feed-water to be used again.

It goes without saying that condenser systems need a constant, ample supply of cooling water and this is supplied in a separate circuit from the cooling tower which cools the condenser cooling water by direct contact with the air and evaporation of a portion of the cooling water in an open tower. Water vapour seen billowing from power plants is evaporating cooling water, not the working fluid.

Back-Pressure Turbines, often used for electricity generation in process industries, do not use condensers. Also called Atmospheric or Non- Condensing Turbines, they do not waste the energy in the steam emerging from the turbine exhaust however, instead it is diverted for use in applications requiring large amounts of heat such as refineries, pulp and paper plants, desalination plants and district heating units. These industries may also use the available steam to power mechanical drives for pumps, fans and materials handling. The boiler and turbine must of course be oversized for the electrical load in order to compensate for the power diverted for other uses.

Practical Machines

Steam turbines come in many configurations. Large machines are usually built with multiple stages to maximise the energy transfer from the steam.

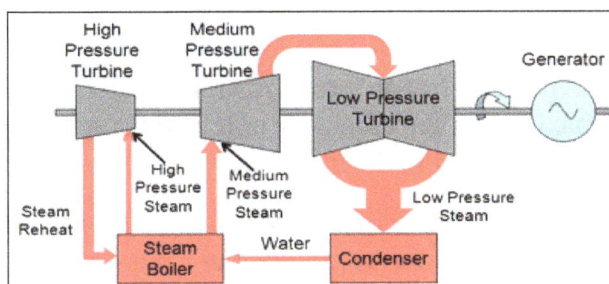

Multi-stage steam turbine generator

To reduce axial forces on the turbine rotor bearings the steam may be fed into the turbine at the mid-point along the shaft so that it flows in opposite directions towards each end of the shaft thus balancing the axial load. The output steam is fed through a cooling tower through which cooling water is passed to condense the steam back to water.

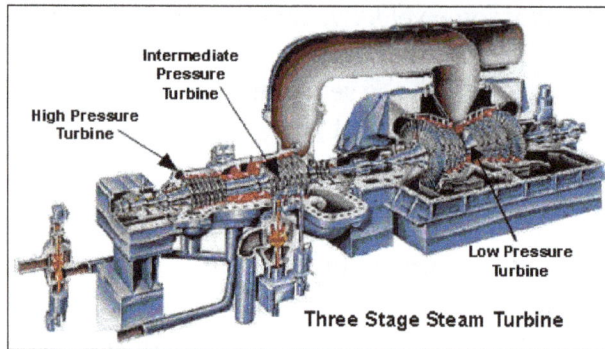

Three Stage Steam Turbine

Turbine power outputs of 1000MW or more are typical for electricity generating plants.

The Steam Turbine as a Heat Engine

Steam turbine systems are essentially heat engines for converting heat energy into mechanical energy by alternately vaporising and condensing a working fluid in a process in a closed system known as the Rankine cycle. This is a reversible thermodynamic cycle in which heat is applied to a working fluid in an evaporator, first to vaporise it, then to increase its temperature and pressure. The high temperature vapour is then fed through a heat engine, in this case a turbine, where it imparts its energy to the rotor blades causing the rotor to turn due to the expansion of the vapour as its pressure and temperature drops. The vapour leaving the turbine is then condensed and pumped back in liquid form as feed to the evaporator.

In this case the working fluid is water and the vapour is steam but the principle applies to other working fluids such as ammonia which may be used in low temperature applications such as geothermal systems. The working fluid in a Rankine cycle thus follows a closed loop and is re-used constantly.

The efficiency of a heat engine is determined only by the temperature difference of the working fluid between the input and output of the engine.

Carnot showed that the maximum efficiency available = $1 - T_c / T_h$ where T_h is the temperature in degrees Kelvin of the working fluid in its hottest state (after heat has been applied) and T_c is its temperature in its coldest state (after the heat has been removed).

To maximise efficiencies, the temperature of the steam fed to the turbine can be as high as 900°C, while a condenser is used at the output of the turbine to reduce the temperature and pressure of the steam to as low a value as possible by converting it back to water. The condenser is an essential component necessary for maximising the efficiency of the steam engine by maximising the temperature difference of the working fluid in the machine.

Using for a typical steam turbine system with an input steam temperature of 543 °C (816K) and a temperature of the condensed water of 23 °C (296K), the maximum theoretical efficiency can be calculated as follows:

Carnot efficiency = (816 - 296)/816 = 64%

But this does not take account of heat, friction and pressure losses in the system. A more realistic value for the efficiency of the steam turbine would be about 50%. Thus, the heat engine is responsible for most of the system energy conversion losses.

This only includes the conversion of the heat energy in the steam to mechanical energy on the turbine shaft. It does not include the efficiency loss in the combustion chamber and boiler in converting the chemical energy of the fuel to heat energy in the steam nor does it include the efficiency losses incurred in the generator if the turbine is used to generate electricity. Taking these losses into account, the overall efficiency of converting the fuel's chemical energy in coal and oil fired plants to electrical energy is typically around 33%.

Electromechanical Energy Transfer (Generator)

The steam turbine drives a generator, to convert the mechanical energy into electrical energy. Typically this will be a rotating field synchronous machine.

The energy conversion efficiency of these high capacity generators can be as high as 98% or 99% for a very large machine.

Ancillary Systems

Apart from the basic steam raising and electricity generating plant, there are several essential automatic control and ancillary systems which are necessary to keep the plant operating safely at its optimum capacity. These include:

- Matching the power output to the demand. Current controls,
- Maintaining the system voltage and frequency,
- Keeping the plant components within their operating pressure, temperature and speed limits,
- Lubrication systems,
- Feeding the fuel to the combustion chamber and removing the ash,
- Pumps and fans for water and air flow,
- Pollution. control - Separating harmful products from the combustion exhaust emissions,
- Cooling the generator,
- Electricity transmission equipment. Transformers and high voltage switching,
- Overload protection, emergency shut down and load shedding.

Nuclear Power Generation

Nuclear plants are different to energy plants such as coal and natural gas, because despite being a thermal generation process they do not need to burn anything to create steam.

In a nuclear plant, uranium atoms are split in a process called fission, which requires low-enriched uranium fuel. Uranium fuel is formed into pellets, one of which can produce as much energy as one tonne of coal, three barrels of oil or 17,000 cubic feet of natural gas. These pellets are generally stacked into 12-foot metal fuel rods, which are grouped together in bundles that are called fuel assemblies.

In nuclear fission, heat and neutrons are released from uranium as the atoms split. The neutrons hit other uranium atoms causing them to split, continuing the cycle. Meanwhile, the released heat causes water within the reactor to boil, which in turn creates the steam that powers the turbines, which powers the generators to make electricity.

Types of Nuclear Reactors

There are two standard types of nuclear reactor, firstly boiling water reactors (BWR), which simply heat up water until it boils to spin turbines and generate electricity. Secondly, pressurised water reactors (PWR), which heat up water to close to boiling point before this water is pumped into a separate supply of water. In this compartment, it becomes steam that is used to powers a turbine.

Disadvantages and Advantages of Nuclear Power Generation

Disadvantages

1. Nuclear power is a controversial method of producing electricity. Many people and environmental organisations are very concerned about the radioactive fuel it needs.

2. There have been serious accidents with a small number of nuclear power stations. The accident at Chernobyl (Ukraine) in 1986, led to 30 people being killed and over 100,000 people being evacuated. In the preceding years another 20,000 people were resettled away from the radioactive area. Radiation was even detected over a thousand miles away in the UK as a result of the Chernobyl accident. It has been suggested that over time 2500 people died as a result of the accident.

3. Some of the waste remains radioactive (dangerous) for thousands of years and is currently stored in places such as deep caves and mines.

4. Storing and monitoring the radioactive waste material for thousands of years has a high cost.

5. Nuclear powered ships and submarines pose a danger to marine life and the environment. Old vessels can leak radiation if they are not maintained properly or if they are dismantled carelessly at the end of their working lives.

6. Many people living near to nuclear power stations or waste storage depots are concerned about nuclear accidents and radioactive leaks. Some fear that living in these areas can damage their health, especially the health of young children.

7. Many Governments fear that unstable countries that develop nuclear power may also develop nuclear weapons and even use them.

Advantages

1. The amount of electricity produced in a nuclear power station is equivalent to that produced by a fossil fuelled power station.

2. Nuclear power stations do not burn fossil fuels to produce electricity and consequently they do not produce damaging, polluting gases.

3. Many supporters of nuclear power production say that this type of power is environmentally friendly and clean. In a world that faces global warming they suggest that increasing the use of nuclear power is the only way of protecting the environment and preventing catastrophic climate change.

4. Many developed countries such as the USA and the UK no longer want to rely on oil and gas imported from the Middle East, a politically unstable part of the world.

5. Countries such as France produce approximately 90 percent of their electricity from nuclear power and lead the world in nuclear power generating technology - proving that nuclear power is an economic alternative to fossil fuel power stations.

6. Nuclear reactors can be manufactured small enough to power ships and submarines. If this was extended beyond military vessels, the number of oil burning vessels would be reduced and consequently pollution.

Electrical Power Distribution

Electrical power distribution is the final stage of an electrical power system, which entails the delivery of electricity to the load. The primary role of electrical power distribution is to carry the electricity from the transmission lines to the loads in the individual customers to the different strata of society.

Electrical Power Distribution System

Electric Power Distribution System states that part of power system which distributes electric power for local use is known as distribution system.

The electrical energy produced at the generating station is conveyed to the consumers through a network of transmission and distribution systems. It is often difficult to draw a line between the transmission and distribution systems of a large power system. It is impossible to distinguish the two merely by their voltage because what was considered as a high voltage a few years ago is now considered as a low voltage. In general, Electric Power Distribution System is that part of power system which distributes power to the consumers for utilization.

The transmission and distribution systems are similar to man's circulatory system. The transmission system may be compared with arteries in the human body and distribution system with capillaries. They serve the same purpose of supplying the ultimate consumer in the city with the life-giving blood of civilization electricity.

Classification of Distribution Systems

A distribution system may be classified on the bases of their properties.

1. Nature of current: According to nature of current, distribution system may be classified as:

- D.C. distribution system
- A.C. distribution system

Now-a-days, A.C. system is universally adopted for distribution of electric power as it is simpler and more economical than direct current method.

2. Type of construction: According to type of construction, distribution system may be classified as:

- Overhead system
- Underground system

The overhead system is generally employed for distribution as it is 5 to 10 times cheaper than the equivalent underground system. In general, the underground system is used at places where overhead construction is impracticable or prohibited by the local laws.

3. Scheme of connection: According to scheme of connection, the Electric Power Distribution System may be classified as:

- Radial system
- Ring main system
- Inter-connected system

Distribution of electric power is done by distribution networks. Distribution networks consist of following main parts:

- Distribution substation,
- Primary distribution feeder,
- Distribution Transformer,
- Distributors,
- Service mains.

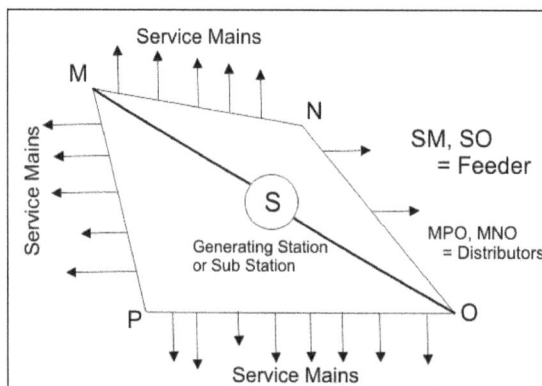

The transmitted electric power is stepped down in substations, for primary distribution purpose. Now these stepped down electric power is fed to the distribution transformerthrough primary distribution feeders. Overhead primary distribution feeders are supported by mainly supporting iron pole (preferably rail pole). The conductors are strand aluminum conductors and they are mounted on the arms of the pole by means of pin insulators. Sometimes in congested places, underground cables may also be used for primary distribution purposes.

Distribution transformers are mainly 3 phase pole mounted type. The secondary of the transformer is connected to distributors. Different consumers are fed electric power by means of the service mains. These service mains are tapped from different points of distributors. The distributors can also be re-categorized by distributors and sub-distributors. Distributors are directly connected to the secondary of distribution transformers whereas sub-distributors are tapped from distributors.

Service mains of the consumers may be either connected to the distributors or sub-distributors depending upon the position and agreement of consumers. Both feeder and distributor carry the electrical load, but they have one basic difference. Feeder feeds power from one point to another without being tapped from any intermediate point. As because there is no tapping point in between, the current at sending end is equal to that of receiving-end of the conductor. The distributors are tapped at different points for feeding different consumers, and hence the current varies along their entire length.

Radial Electrical Power Distribution System

In the early days of electrical power distribution system, different feeders radially came out from the substation and connected to the primary of distribution transformer.

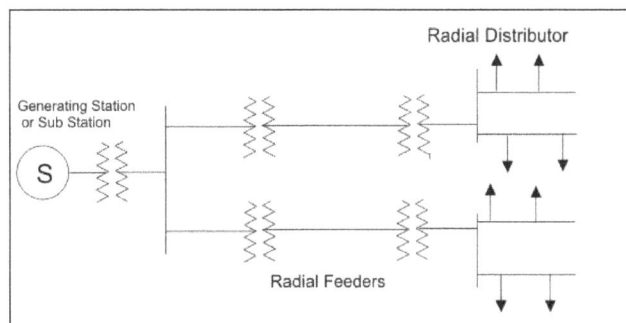

But radial electrical power distribution system has one major drawback that in case of any feeder failure, the associated consumers would not get any power as there was no alternative path to feed the transformer. In case of transformer failure also, the power supply is interrupted. In other words, the consumer in the radial electrical distribution system would be in darkness until the feeder or transformer was rectified.

Ring Main Electrical Power Distribution System

The drawback of radial electrical power distribution system can be overcome by introducing a ring main electrical power distribution system. Here one ring network of distributors is fed by more than one feeder. In this case, if one feeder is under fault or maintenance, the ring distributor is still energized by other feeders connected to it. In this way, the supply to the consumers is not affected even when any feeder becomes out of service. In addition to that, the ring main system is also provided with different section isolates at different suitable points. If any fault occurs on any part, of the ring, this part can easily be isolated by opening the associated section isolators on both sides of the faulty zone transformer directly.

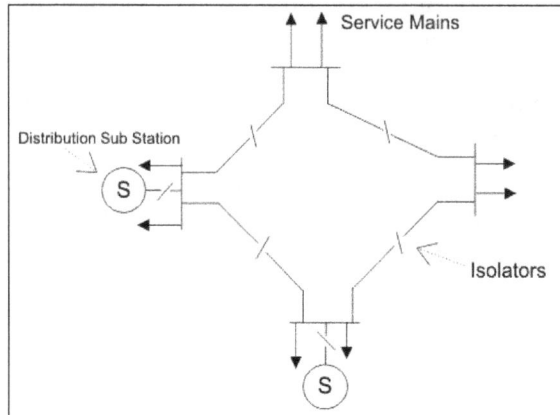

In this way, supply to the consumers connected to the healthy zone of the ring can easily be maintained even when one section of the ring is under the shutdown. The number of feeders connected to the ring main electrical power distribution system depends upon the following factors:

1. Maximum Demand of the System: If it is more, then more numbers of feeders feed the ring.

2. Total Length of the Ring Main Distributors: It length is more, to compensate the voltage drop in the line, more feeders to be connected to the ring system.

3. Required Voltage Regulation: The number of feeders connected to the ring also depends upon the permissible allowable, voltage drop of the line.

AC Power Distribution System

According to phases and wires involved, an AC distribution system can be classified as:

1. Single phase, 2-wire system,

2. Single phase, 3-wire system,

3. Two phase, 3-wire system,

4. Two phase, 4-wire system,

5. Three phase, 3-wire system,

6. Three phase, 4-wire system.

Single Phase and 2-Wire Distribution

This system may be used for very short distances. The following figure shows a single phase two wire system with – figure (a) one of the two wires earthed and figure (b) mid-point of the phase winding is earthed.

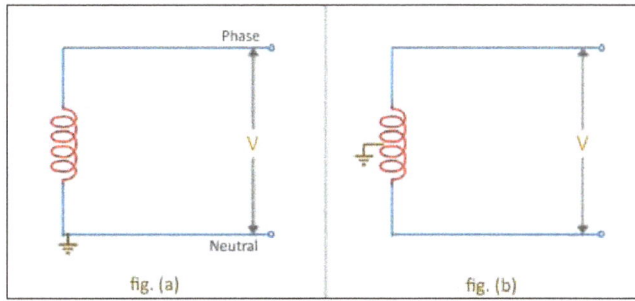

fig. (a)   fig. (b)

Single Phase and 3-Wire System

This system is identical in principle with 3-wire dc distribution system. The neutral wire is center-tapped from the secondary winding of the transformer and earthed. This system is also called as split-phase electricity distribution system.

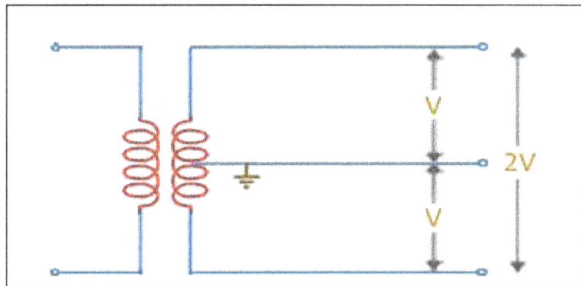

Two Phase and 3-Wire System

In this system, the neutral wire is taken from the junction of two phase windings whose voltages are in quadrature with each other. The voltage between neutral wire and either of the outer phase wires is V. Whereas, the voltage between outer phase wires is √2V. As compared to a two-phase 4-wire system, this system suffers from voltage imbalance due to unsymmetrical voltage in the neutral.

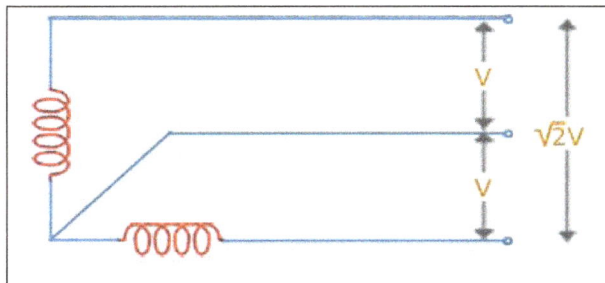

Two Phase and 4-Wire System

In this system, 4 wires are taken from two phase windings whose voltages are in quadrature with each other. Mid-point of both phase windings are connected together. If the voltage between the two wires of a same phase is V, then the voltage between two wires of different phase would be 0.707V.

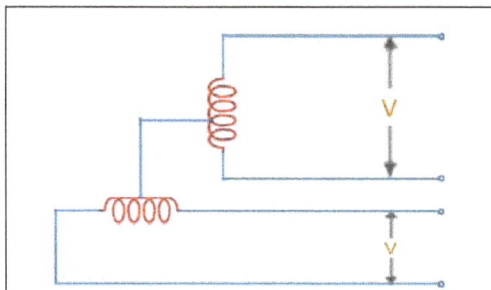

Three Phase and 3-Wire Distribution System

Three phase systems are very widely used for AC power distribution. The three phases may be delta connected or star connected with star point usually grounded. The voltage between two phases or lines for delta connection is V, where V is the voltage across a phase winding. For star connection, the voltage between two phases is $\sqrt{3}$V.

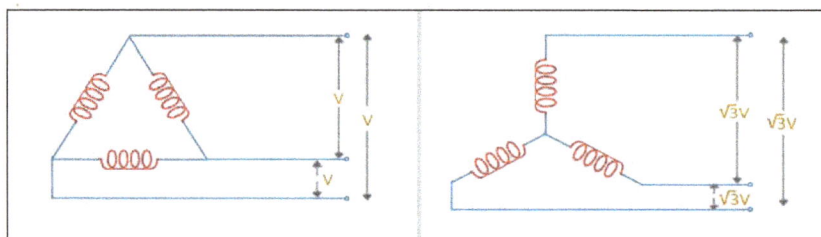

Three Phase and 4-Wire Distribution System

This system uses star connected phase windings and the fourth wire or neutral wire is taken from the star point. If the voltage of each winding is V, then the line-to-line voltage (line voltage) is $\sqrt{3}$V and the line-to-neutral voltage (phase voltage) is V. This type of distribution system is widely used in India and many other countries. In these countries, standard phase voltage is 230 volts and line voltage is $\sqrt{3} \times 230$ = 400 volts. Single phase residential loads, single phase motors which run on 230 volts etc. are connected between any one phase and the neutral. Three phase loads like three-phase induction motors are put across all the three phases and the neutral.

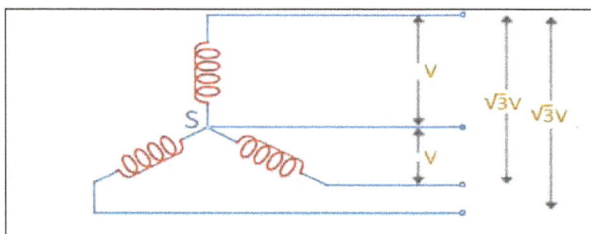

DC Power Distribution System

Electrical power is almost exclusively generated, transmitted and distributed in AC form. However, for certain applications, DC supply is absolutely necessary. For example, variable speed machinery incorporating DC motors, critical areas where storage battery reserves are necessary. Following are some points that make us think about dc power distribution.

- Advancements in Power electronics have made it possible to transform DC voltage levels and conversion between AC and DC efficiently. It is now possible to replace existing AC distribution network with DC distribution network.

- Distribution generation from solar and wind energy is increasing rapidly and both of these sources are intrinsically DC.

- A large number of office and household appliances internally require low voltage DC. These appliances are fed with AC supply and then transformed to lower voltage and converted into DC by an internal circuitry.

- Harmonic issues, phase balancing problems, skin effect etc. are not present in DC systems.

- DC energy can be stored easily in batteries and fuel cells. Such backup batteries can be utilized easily in case of supply failure.

Types of DC Power Distribution

Wherever DC power distribution is required, AC power from the transmission network can be rectified at a substation using converting equipment and then fed to the dc distribution system. AC consumers can also be connected to DC system using a DC to AC inverter. A low voltage DC distribution system is of two types.

Unipolar DC Distribution System (2-Wire DC System)

As the name suggests, this system uses two conductors, one is positive conductor and the other one is negative conductor. The energy is transmitted at only one voltage level to all the consumers using this system. A typical unipolar dc power distribution system is as shown in the following figure.

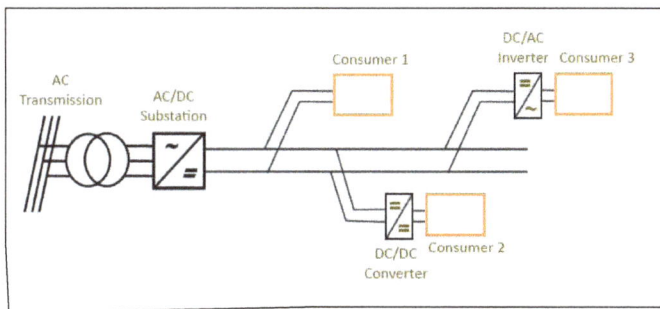

Bipolar DC Distribution System (3-Wire DC System)

This is basically a combination of two series connected unipolar DC systems. It consists of three conductors, two outer conductors (one is positive and the other is negative) and one middle conductor which acts as neutral. This system leaves following connection choices to a consumer:

- Between positive conductor and neutral,

- Between negative conductor and neutral,

- Between positive and negative conductor (double voltage),

- Positive to negative with neutral connected.

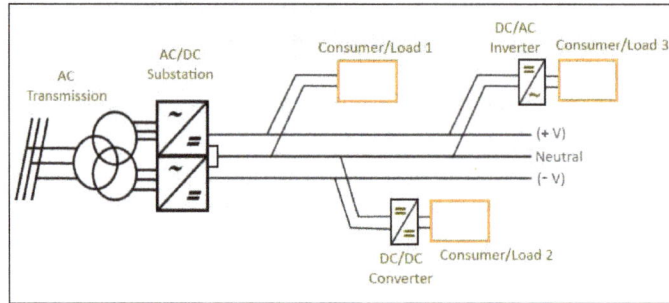

The above figures of unipolar and bipolar dc distribution system suggest that, DC to DC converter or DC to AC inverter can be installed at the consumer's premises according to consumer's or load's requirement. Consumers can also be directly connected to the DC distributors if the distribution voltage level is similar as per their requirement.

Types of DC Distributors

DC distributors are usually classified on the basis of the way they are fed by the feeders. Following are the four types of DC distributors:

- Distributor fed at one end,

- Distributor fed at both ends,

- Distributor fed at center,

- Ring distributor.

Distributor Fed at one end

In this type, distributor is connected to the supply at one end and loads are tapped at different points along its length. The following figure shows the single line diagram of a distributor fed at one end. It worth to note that:

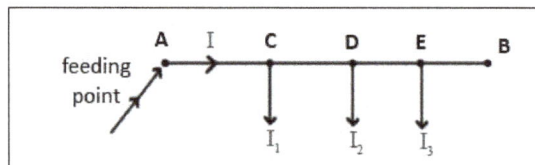

- The current in various sections of the distributor away from the feeding point goes on decreasing. From the above figure, the current in section DE is less than the current in section CD and likewise.

- The voltage also goes on decreasing away from the feeding point. In the above figure, voltage at point E will be minimum.

- In case of a fault in any section of the distributor, the whole distributor will have to be disconnected from the supply. Thus, continuity of supply is interrupted.

Distributor Fed at both Ends

In this type, the distributor is connected to supply at both ends and voltages at feeding points may or may not be equal. The minimum voltage occurs at some load point which is shifted with the variation of load on different sections of the distributor.

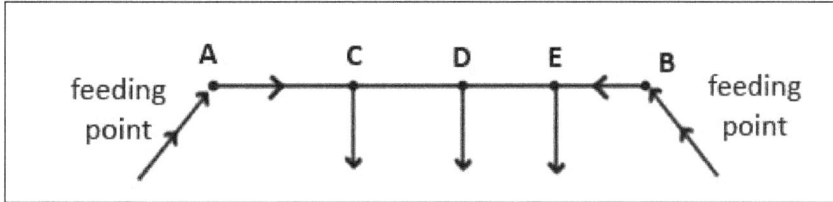

- If a fault occurs at any feeding point, continuity of the supply is ensured from the other feeding point.

- If a fault occurs on any section of the distributor, continuity of the supply is ensured on both sides of the fault with respective feeding points.

- The conductor cross-section area required for a doubly fed distributor is much less than that required for a distributor fed at one end.

Distributor Fed at the Center

As the name implies, the distributor is supplied at the center point. Voltage drop at the farthest ends is not as large as that would be in a distributor fed at one end.

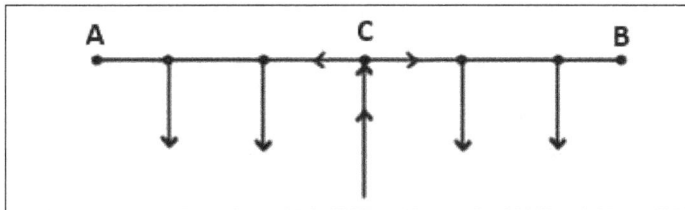

Ring Main DC Distributor

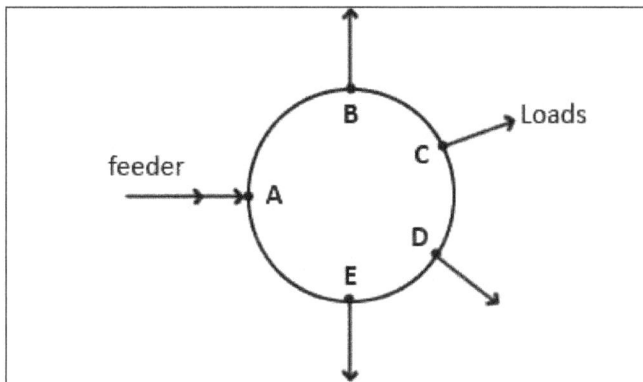

In this type, the distributor is in the form of a closed ring and fed at one point. This is equivalent to a straight distributor fed at both ends with equal voltages.

References

- Power-generation-safety-topics: nttinc.com, Retrieved 17 May, 2019

- Spotlight-on-power-generation: aplant.com, Retrieved 19 February, 2019

- Solar-energy-power-generation: electronicshub.org, Retrieved 21 June, 2019

- Wind-power: mpoweruk.com, Retrieved 24 April, 2019

- Hydro-power: mpoweruk.com, Retrieved 7 July, 2019

- Re-geo-elec-production: nrel.gov, Retrieved 28 January, 2019

- Geothermal-power-plants: energyeducation.ca, Retrieved 2 August, 2019

- Working-principle-of-steam-turbine, classification-or-types-of-steam-turbine: mechanicaltutorial.com, Retrieved 22 June, 2019

- Steam-turbines: mpoweruk.com, Retrieved 24 March, 2019

- What-is-a-nuclear-power-station: power-technology.com, Retrieved 4 July, 2019

- Nuclear, energy: technologystudent.com, Retrieved 19 January, 2019

- Electric-power-distribution-system: eeeguide.com, Retrieved 9 May, 2019

- Types-of-ac-power-distribution-systems: electricaleasy.com, Retrieved 28 February, 2019

- Dc-power-distribution-systems: electricaleasy.com, Retrieved 17 August, 2019

6

Electric Power Quality and Control

Electric power quality deals with the voltage, waveform and frequency of electricity. The process through which data on power quality is gathered, analyzed and interpreted is known as power quality monitoring. This chapter has been carefully written to provide an easy understanding of the varied facets of electric power quality and its control.

Electric Power Quality

If the power supplied to devices or equipments is deficient, it results in a poor performance. Good power quality makes the equipments function properly without affecting performance or life expectancy.

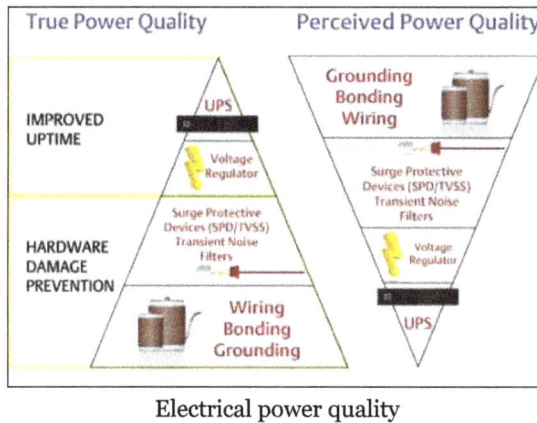

Electrical power quality

IEEE standard defines electrical power quality as "the concept of powering and grounding sensitive electronic equipment in a manner suitable for the equipment with precise wiring system and other connected equipment". It is deviation of voltage and currents from the ideal or actual waveforms.

In the figure, power supplied at mains is pure sine waves of current and voltages. While power reaches to the load it no longer maintains its shape due to non-linear switching devices.

As observed, the shape it deviated from the ideal former one. This deviation causes severe problems in electrical equipments like light flickering, malfunction of various devices, low motor speed running, etc.

By using power quality analyzers we can estimate or analyze the distorted waveform.

Electrical power quality plays an important role in supplying electricity effectively to the consumers. As power becomes more essential and valuable resource for the entire world, it is important to maintain its quality at all levels of usage for reliable working of equipments.

Due to usage of non-linear loads and power electronic equipments in power system transmission, distribution and utilization sectors leads to distortion in voltage and current waveforms. We are already aware of the total harmonic distortion by phase control and integral control of AC power.

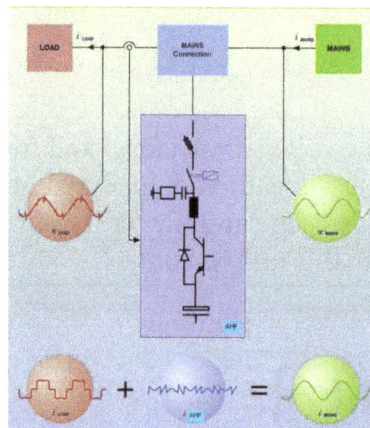

Deviation of waveforms from actual

Now a day's power distribution companies are showing competitive nature to improve power quality by increasing concern over it to get the profitability and customer satisfaction.

Power Quality Issues

Quality of the power is decided by the end users. If the power equipment works satisfactorily for given supply then power is at good quality. If it doesn't functions well or fails to work, then power quality is bad. Reasons for bad power quality or power quality issues are:

Power Frequency Disturbances

1. Voltage sags and swells: Voltage sag or dip is the decrease of voltage levels from nominal values at power frequency. It lasts from about half of a cycle to several seconds. Low voltages are due to several factors like electrical motors, arc furnaces, utility problems, flickering etc.

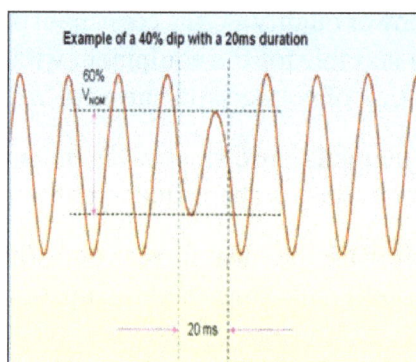

Voltage sags

Motors like different types of induction motors during starting take very large current, which results in a drastic voltage drop.

Also arc furnaces initially take large amperes to produce high temperatures. Utilities drops the voltages by some of factors like lightning, contact of trees, birds and animals to power supply lines, switching operations, insulation failures, etc.

Voltage swells occur due to transfer of loads from one source to another, sudden rejection and application loads. Flickering is a low frequency problem that occurs mainly at starting or low voltage conditions.

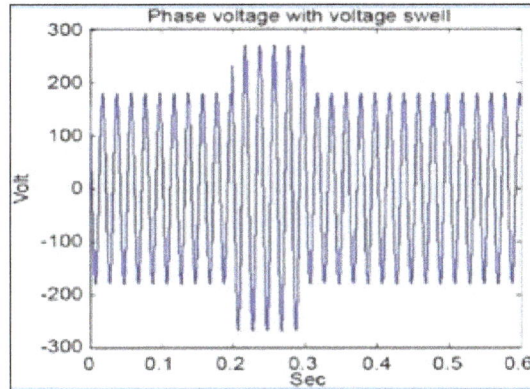

Voltage swells

Flickering is due to low voltages or frequency that can be observed by the human eye.

Voltage sags and swells results in malfunction of equipments, loss of efficiency of motors, insulation failures, fluctuation of light illumination, tripping of relays and contractors, etc.

Power frequency disturbances are not easily cured if they arise at source level because it deals with high powers. However these can be reduced if occurred internally due to loads by separating off end loads from the sensitive loads.

2. Electrical Transients: Transients are sub-cycle disturbances that last for less than one cycle of AC waveforms. Due to limited frequency response or sampling rate, detection and measurement of transients are very difficult.

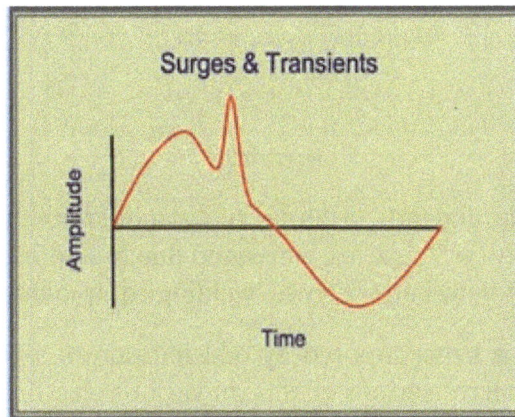

Electrical transients

These are also sometimes called as spikes, surges, power pulses, etc. These occur due to atmospheric disturbances like lighting and solar flares, fault current interruptions, switching the loads, switching capacitor banks, switching power lines, etc.

Some of the devices are designed with transients in mind but most of the devices can handle few transients depends on severity of the transient and life of equipment. These transients are limited by surge protection suppressors, filters and other transient suppressors as shown in figure.

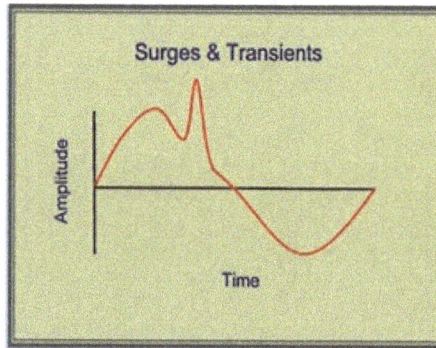

Electrical transient's suppression

3. Harmonics: Harmonic nature of voltage and currents is the deviation from the original or pure sine waves. Harmonic frequencies are integral multiples of fundamental frequency and are very common in electric power systems.

Order of harmonics differentiates these as even (2, 4, 6, 8, 10) and odd types (3, 5, 7, 9, 11). Major nonlinear loads produce odd harmonics and even harmonics are produced due to uneven operations of the electrical devices like transformer magnetizing currents contains the even harmonic components.

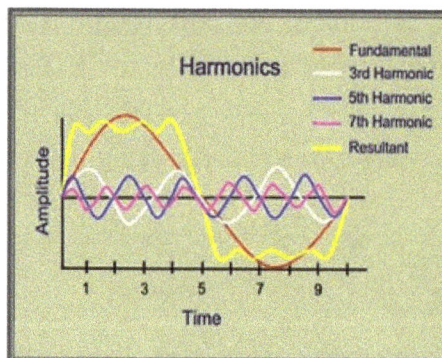

Harmonics

Frequency of these harmonics depends on order of harmonics as 2nd harmonic frequency is 2 times the fundamental frequency. These are generated due to nonlinear loads, arc furnaces, electric motors, UPS systems, different battery types, welding equipments, etc.

The fundamental waveform is superimposed by odd harmonics, which results in the distorted waveforms. These harmonics have serious effects on various electrical equipments such as overheating of cables and equipments, interference with communication lines, errors while indicating electrical parameters, probability to produce resonant conditions, etc.

These can be easily measured by harmonic analyzers and reduced by using various harmonic filters like active and passive types.

Power Factor

Power factor is another main factor which affects the electrical power quality. Low power factor causes several problems like overheating of motors and poor lightening. It also leads to the users being penalized to meet electric demands. Power factor is the ratio of active power to apparent power and determines the amount of electrical power utilization.

Suppose if power factor is 0.8, tells that 80 percent of the power is utilized and remaining energy is wasted as losses. Low power factor is due to induction motors, apparent power elements in electrical power system network, etc.

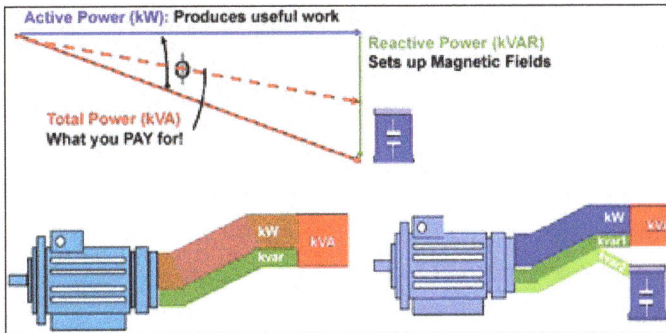

Low power factor is improved by using power factor correction devices such as capacitor filter banks, synchronous condensers and other compensation equipments.

Power factor improvement, with use of capacitors, results in a reduction of electric bills. Here apparent power drawn from the supply is reduced by capacitors which offer leading power in nature.

Grounding

Good power quality includes safety to the appliances as well as to operators. Grounding provides system protection as well as equipment protection. Earth serves as constant reference potential with other potential which is going to be measured.

If the equipment body is not properly grounded it results in severe shock to individuals. System ground protects various equipments against fault conditions and other abnormal conditions occurring at electrical power systems.

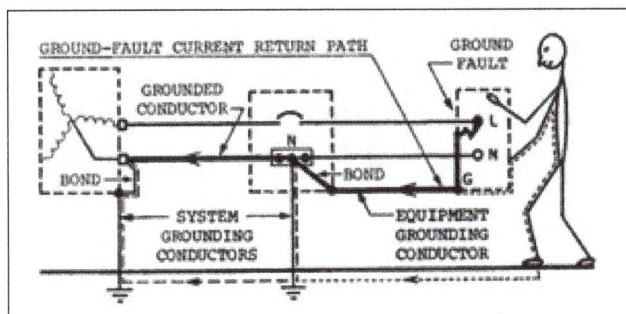

Equipment and system groundings

Signal reference ground is entirely different from normal grounding since it does not provide any protection to equipment or individuals. But it is necessary for proper working of electronic components or devices to provide low impedance path or reference.

Power Quality Monitoring

Power quality monitoring is the process of gathering, analyzing, and interpreting raw measurement data into useful information. The process of gathering data is usually carried out by continuous measurement of voltage and current over an extended period. The process of analysis and interpretation has been traditionally performed manually, but recent advances in signal processing and artificial intelligence fields have made it possible to design and implement intelligent systems to automatically analyze and interpret raw data into useful information with minimum human intervention.

Power quality monitoring programs are often driven by the demand for improving the system wide power quality performance. Many industrial and commercial customers have equipment that is sensitive to power disturbances, and, therefore, it is more important to understand the quality of power being provided. Examples of these facilities include computer networking and telecommunication facilities, semiconductor and electronics manufacturing facilities, biotechnology and pharmaceutical laboratories, and financial data-processing centers.

Monitoring Considerations

The monitoring objectives often determine the choice of monitoring equipment, triggering thresholds, methods for data acquisition and storage, and analysis and interpretation requirements. Several common objectives of power quality monitoring are summarized here.

Monitoring to Characterize System Performance

This is the most general requirement. A power producer may find this objective important if it has the need to understand its system performance and then match that system performance with the needs of customers. System characterization is a proactive approach to power quality monitoring. By understanding the normal power quality performance of a system, a provider can quickly identify problems and can offer information to its customers to help them match their sensitive equipment's characteristics with realistic power quality characteristics.

Monitoring to Characterize Specific Problems

Many power quality service departments or plant managers solve problems by performing short-term monitoring at specific customer sites or at difficult loads. This is a reactive mode of power quality monitoring, but it frequently identifies the cause of equipment incompatibility, which is the first step to a solution.

Monitoring as part of an Enhanced Power Quality Service

Many power producers are currently considering additional services to offer customers. One of these services would be to offer differentiated levels of power quality to match the needs of specific customers. A provider and customer can together achieve this goal by modifying the power system or by installing equipment within the customer's premises. In either case, monitoring becomes essential to establish the benchmarks for the differentiated service and to verify that the utility achieves contracted levels of power quality.

Monitoring as part of Predictive or Just-in-time Maintenance

Power quality data gathered over time can be analyzed to provide information relating to specific equipment performance. For example, a repetitive arcing fault from an underground cable may signify impending cable failure, or repetitive capacitor-switching restrikes may signify impending failure on the capacitor-switching device. Equipment maintenance can be quickly ordered to avoid catastrophic failure, thus preventing major power quality disturbances which ultimately will impact overall power quality performance.

The monitoring program must be designed based on the appropriate objectives, and it must make the information available in a convenient form and in a timely manner (i.e., immediately). The most comprehensive monitoring approach will be a permanently installed monitoring system with automatic collection of information about steady-state power quality conditions and energy use as well as disturbances.

Power Management System

An electrical power management system (EPMS) is an electronic system that provides fine-grained information about the flow of power in an electrical power generation system or power substation.

EPMS record and provide data about power systems and power-related events. That information is used to manage power generation efficiencies, batteries and capacitor banks, gas or steam turbine relays and other systems in power generation stations and power substations. EPMS can visually display real-time or historical data. The system ties together the essential data that formerly had to be checked on numerous readouts and gauges by equipment operators. Supervisory control and data acquisition systems (SCADA) systems often use EPMS, especially those used in power plants.

Besides power generation stations, EPMS can be found in manufacturing plants, on large ships' generators and in similar high power demand locations. Some EPMS are their own systems, while others integrate with supervisory control and data acquisition (SCADA) and yet others are hybrid systems.

EPMS that include generator protection and control (GPC) relays and those that are integrated with SCADA can automate many power-related relays. This control helps increase power efficiency, especially in times of high draw. Some products claim they can help reduce peak power draw by 50%. Applied to the power grid, this reduction could theoretically alleviate concerns of a power crisis resulting from peak demand.

Better power management is helpful in terms of smoothing power demands. Smoothing out peak and low demand is often very beneficial and lower in cost as the problem in energy systems is often not that overall average power is too high but that peak draw times exceed momentary power production.

SCADA

SCADA stands for supervisory control and data acquisition. It is a type of software application program for process control. SCADA is a central control system which consist of controllers network interfaces, input/output, communication equipments and software. SCADA systems are used to monitor and control the equipments in the industrial process which include manufacturing, production, development and fabrication. The infrastructural processes include gas and oil distribution, electrical power, water distribution. Public utilities include bus traffic system, airport. The SCADA system takes the reading of the meters and checks the status of sensors in regular interval so that it requires minimal interference of human.

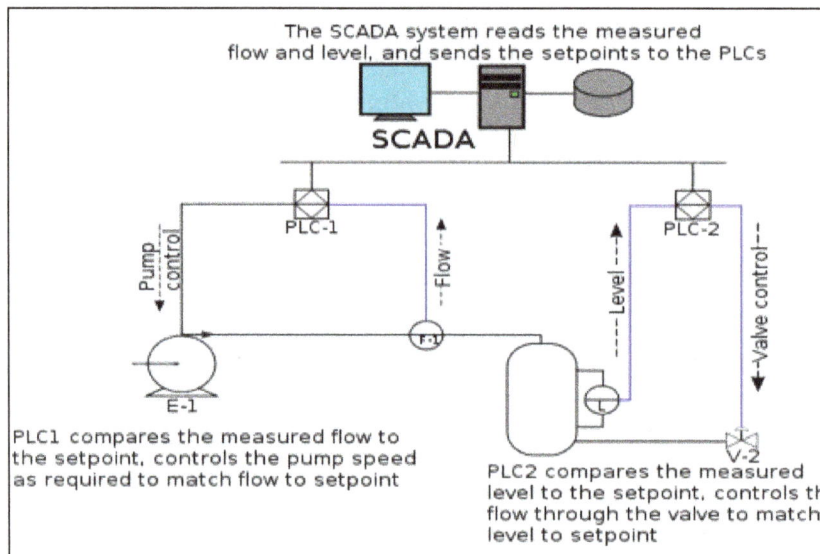

A large number of processes occur in large industrial establishment. Every process you need to monitor is very complex because each machine gives different output. The SCADA system used to gather the data from sensors and instruments located at remote area. The computer then processes this data and presents in a timely manner. The SCADA system gathers the information (like leak on a pipeline occurred) and transfer the information back to the system while giving the alerts that leakage has occurred and displays the information in a logical and organized fashion. The SCADA system used to run on DOS and UNIX operating systems.

Architecture

Generally SCADA system is a centralized system which monitors and controls entire area. It is purely software package that is positioned on top of hardware. A supervisory system gathers data on the process and sends the commands control to the process. The SCADA is a remote terminal unit which is also known as RTU. Most control actions are automatically performed by RTUs or PLCs. The RTUs consist of programmable logic converter which can be set to specific requirement.

For example, in the thermal power plant the water flow can be set to specific value or it can be changed according to the requirement.

The SCADA system allows operators to change the set point for the flow, and enable alarm conditions in case of loss of flow and high temperature and the condition is displayed and recorded. The SCADA system monitors the overall performance of the loop. The SCADA system is a centralized system to communicate with both wire and wireless technology to Clint devices. The SCADA system controls can run completely all kinds of industrial process.

Example: If too much pressure in building up in a gas pipe line the SCADA system can automatically open a release valve.

Hardware Architecture

The generally SCADA system can be classified into two parts:

- Clint layer,

- Data server layer.

The Clint layer which caters for the man machine interaction. The data server layer which handles most of the process data activities.

The SCADA station refers to the servers and it is composed of a single PC. The data servers communicate with devices in the field through process controllers like PLCs or RTUs. The PLCs are connected to the data servers either directly or via networks or buses. The SCADA system utilizes a WAN and LAN networks, the WAN and LAN consists of internet protocols used for communication between the master station and devices. The physical equipments like sensors connected to the PLCs or RTUs. The RTUs convert the sensor signals to digital data and sends digital data to master. According to the master feedback received by the RTU, it applies the electrical signal to relays. Most of the monitoring and control operations are performed by RTUs or PLCs as we can see in the figure.

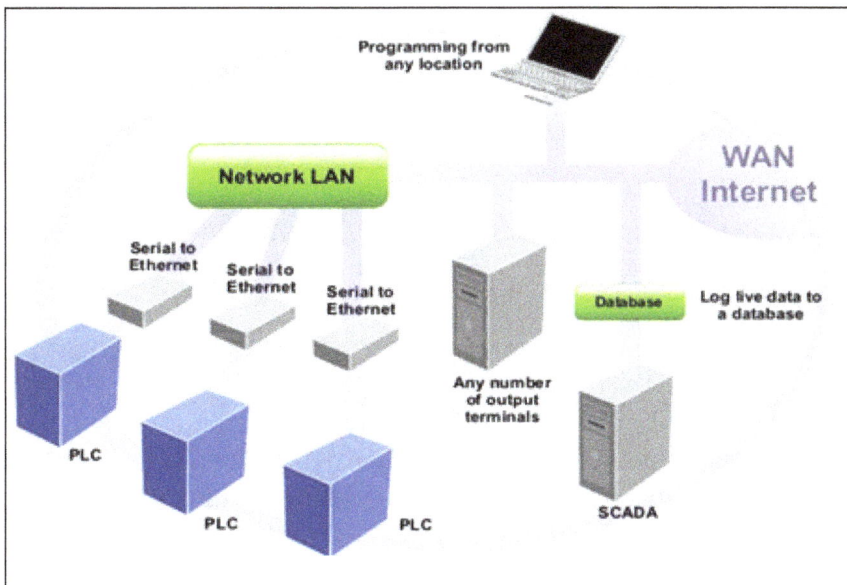

Software Architecture

Most of the servers are used for multitasking and real time database. The servers are responsible for data gathering and handling. The SCADA system consists of a software program to provide trending, diagnostic data, and manage information such as scheduled maintenance procedure, logistic information, detailed schematics for a particular sensor or machine and expert system troubleshooting guides. This means the operator can see a schematic representation of the plant being controlled.

Example: Alarm checking, calculations, logging and archiving; polling controllers on a set of parameter, those are typically connected to the server.

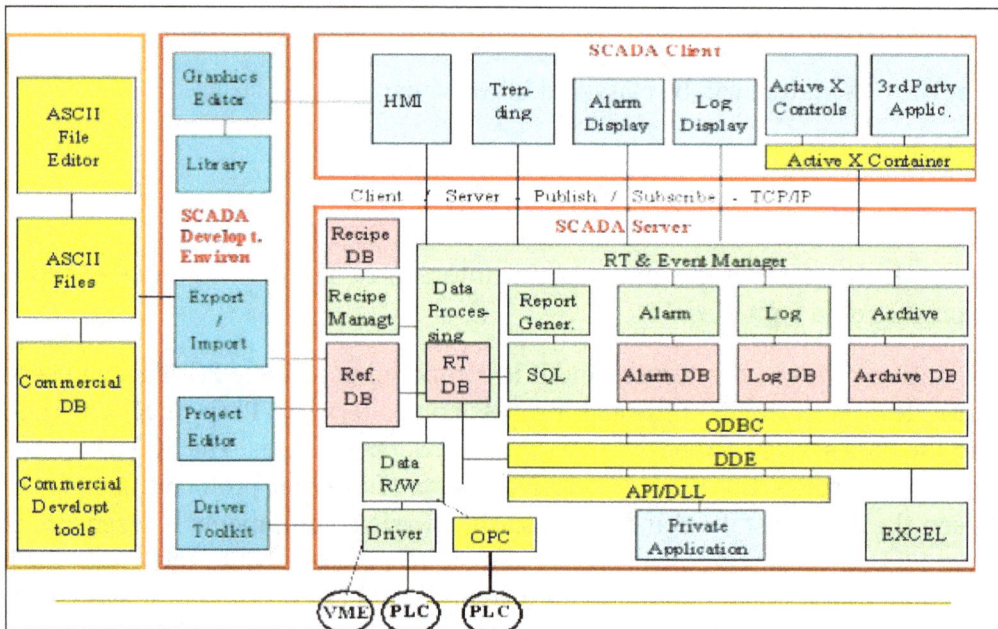

Typical generic software architecture of SCADA systems

Working Procedure of SCADA system

The SCADA system performs the following functions:

- Data Acquisitions

- Data Communication

- Information/Data presentation

- Monitoring/Control

These functions are performed by sensors, RTUs, controller, communication network. The sensors are used to collect the important information and RTUs are used to send this information to controller and display the status of the system. According to the status of the system, the user can give command to other system components. This operation is done by the communication network.

Data Acquisitions

Real time system consists of thousands of components and sensors. It is very important to know the status of particular components and sensors. For example, some sensors measure the water flow from the reservoir to water tank and some sensors measure the value pressure as the water is release from the reservoir.

Data Communication

The SCADA system uses wired network to communicate between user and devices. The real time applications use lot of sensors and components which should be control remotely. The SCADA system uses internet communications. All information is transmitted through internet using specific protocols. Sensor and relays are not able to communicate with the network protocols so RTUs used to communicate sensors and network interface.

Information/Data presentation

The normal circuit networks have some indicators which can be visible to control but in the real time SCADA system, there are thousands of sensors and alarm which are impossible to be handled simultaneously. The SCADA system uses human machine interface (HMI) to provide all of the information gathered from the various sensors.

Human Machine Interface

The SCADA system uses human machine interface. The information is displayed and monitored to be processed by the human. HMI provides the access of multiple control units which can be PLCs and RTUs. The HMI provides the graphical presentation of the system. For example, it provides the graphical picture of the pump connected to the tank. The user can see the flow of the water and pressure of the water. The important part of the HMI is an alarm system which is activated according to the predefined values.

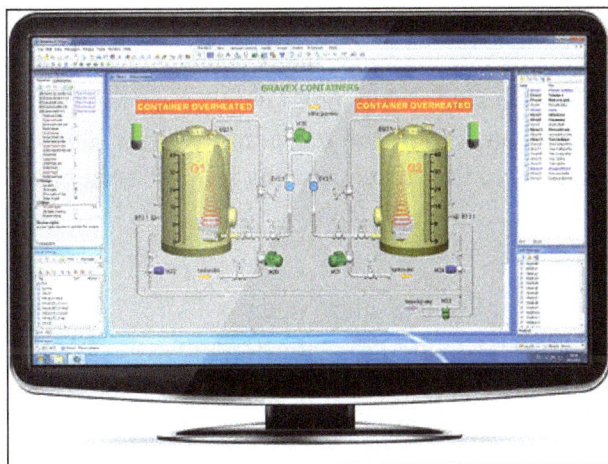

Example: The tank water level alarm is set 60% and 70% values. If the water level reaches above 60% the alarm gives normal warning and if the water level reach above 70% the alarm gives critical warning.

Monitoring/Control

The SCADA system uses different switches to operate each device and displays the status at the control area. Any part of the process can be turned ON/OFF from the control station using these switches. SCADA system is implemented to work automatically without human intervention but at critical situations it is handled by man power.

SCDA for Remote Industrial plant

In large industrial establishments many process occur simultaneously and each needs to be monitored, which is actually a complex task. The SCADA systems are used to monitor and control the equipments in the industrial processes which include water distribution, oil distribution and power distribution. The main aim of this project is to process the real time data and control the large scale remote industrial environment. In the real time scenario, a temperature logging system for a remote plant operation is taken.

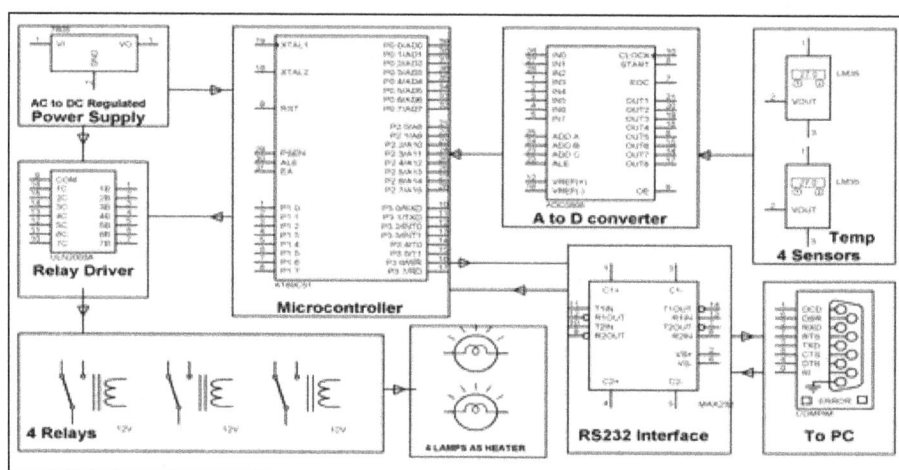

The temperature sensors are connected to the microcontroller, which is connected to the PC at the front end and software is loaded on the computer. The data is collected from the temperature sensors. The temperature sensors continuously send the signal to the microcontroller which accordingly displays these values on its front panel. One can set the parameters like low limit and high limit on the computer screen. When the temperature of a sensor goes above set point the microcontroller send a command to the corresponding relay. The heaters connected through relay contacts are turned OFF and ON.

Applications

- Power generation, transmission and distribution.

- Water distribution and reservoir system.

- Public buildings like electrical heating and cooling system.

- Generators and turbines.

- Traffic light control system.

Advantages

- The SCADA system provides on-board mechanical and graphical information.

- The SCADA system is easily expandable. We can add set of control units and sensors according to the requirement.

- The SCADA system ability to operate critical situations.

References

- Important-factors-affecting-electrical-power-quality: elprocus.com, Retrieved 1 April, 2019

- Power-Quality-Monitoring, electrical-engineering: idc-online.com, Retrieved 3 July, 2019

- Electrical-power-management-system: techtarget.com, Retrieved 9 May, 2019

- Scada-systems-work: elprocus.com, Retrieved 18 March, 2019

7

Applications of Power Electronics

Power electronics are applied in a variety of fields for converting AC current to DC, DC to AC, AC to AC and DC to DC. Some of the fields where power electronics are used are medical science, automotive and electrical engineering. The diverse applications of power electronics in these fields have been thoroughly discussed in this chapter.

Medical Applications

In an era of electronic engineering, we are using electronics for various applications in medical electronics, by that we are able to modify the medical treatment. Medical electronics are most widely developing fields of this era. Medical electronics are finding cures for almost all diseases and to implement treatment. By using medical electronics doctors and surgeons can do medical examinations in a very smart way. Medical electronics provides sophisticated equipment with precision.

Blood Gas Analyser

The best application of electronics in the medical field is gas analyzer. It is used to calculate the pressure of the chemical elements like carbon monoxide, nitrogen, oxygen in blood. By analyzing results we able to understand if any disorder in blood, particularly after we feel sick quite 2 days. By exploiting results we are able to observe if any disorder when we feel sickness within the basic level solely within the home. It is often enforced as medical electronics projects.

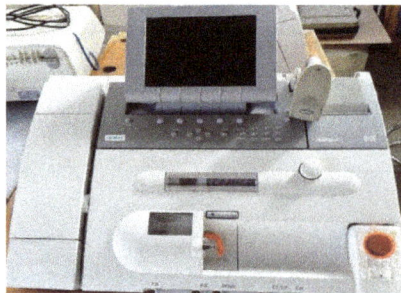

Blood gas analyser

Blood collected from the person is introduced within the chemical device strip that has particle selective electrodes, by exploitation by device amplifiers and analog electronic device; the results are going to be shown in a digital manner with ADC for a microcontroller. Then the output is going to be displayed within the digital display module in terms of millimeters of mercury (mmHg), kilopascals (kPa), typical values for the carbon monoxide and dioxide measure thirty four(34) to thirty five (35) mm Hg , which of Oxygen in between eighty(80) to ninety (90) mm Hg.

Blood Glucose Monitor

Blood glucose monitor is used to calculate glucose level of diabetic patients. These devices are often designed as medical electronic projects. The working of blood glucose monitor is, when a little drop of blood is placed on the chemical strip, the strip has sensors to live content of various chemical components, within seconds it will calculate the amount of glucose in the blood and displays by using a LED display.

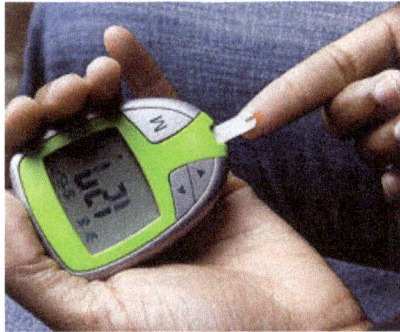

Blood glucose monitor

It is very helpful to observe sugar levels with less prices and straight forward approach. It is a relief to a heap of diabetic patients. Making of Blood Glucose Monitor is easy by usual materials which are available in the market.

Brain Wave Machine

Brain wave machine is one type of instrument in medical electronics which is used to record the electrical activity of the scalp with Electroencephalography by firing of neurons within the brain. It processes the data that has taken from the electrodes which are placed on the scalp and can be displayed within the screen. It is helpful in the treatment of disorders of brains like sleeping disorder, brain death, and mental unhealthiness, also in emergency units at hospitals. These types of electronic devices are used in the medical field in the treatment of mental issues.

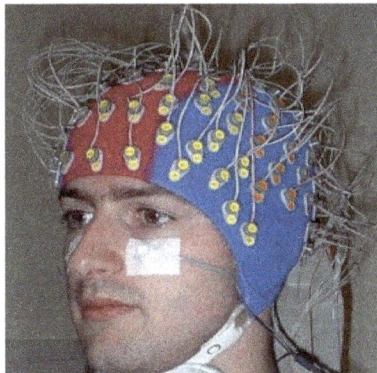

Brain wave machine

Electronic Cardiac Monitor

The Electronic Cardiac Monitor is used in all medical electronics applications. This medical electronic device is used to display the electrical and pressure waveforms of the cardiac system. By inserting specific electrodes on the various elements of the body we can get ECG of the cardiac system. It will observe irregular activity within the cardiac system and heart issues. It is used throughout medical treatment and especially while surgery.

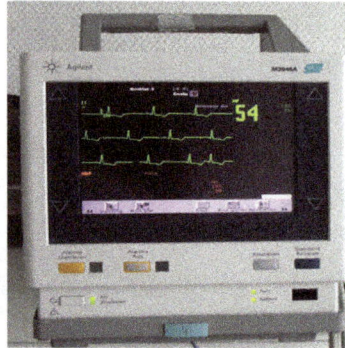
Electronic cardiac monitor

Digital Thermometers

The digital thermometers are used to sense the temperature of the body and these devices are portable, have permanent probes, and a convenient digital display. These devices are used in different industries to control processes in scientific research, the study of weather and in medicine.

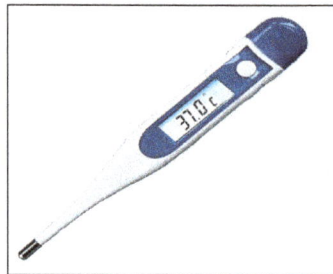
Digital thermometers

IR Thermometers

IR thermometer is used to measure the temperature by detective work with radiation generated by the body. These devices are often used in airports for defectively knowing the condition of passenger's health from a distance to observe the diseases like viral hemorrhagic fever like EBOLA, SAARC etc. This system consists of a lens to focus the infrared (IR) energy onto a target body, and detects the energy, and displayed in the form of electrical signal which will be displayed in units of temperature.

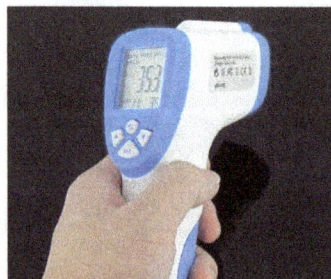
IR Thermometer

Defibrillator

Defibrillator is used in emergency conditions like heart attack occurs. It affects the rhythm of the heart such as ventricular fibrillation, cardiac arrhythmia and pulseless ventricular tachycardia. The working procedure of the Defibrillator involves, when the electric shock delivers to the heart, it causes depolarization of the muscles of the heart and regenerates normal conduction of the electrical pulse of the heat. There are different types of defibrillators include implanted, trans venous and external defibrillators.

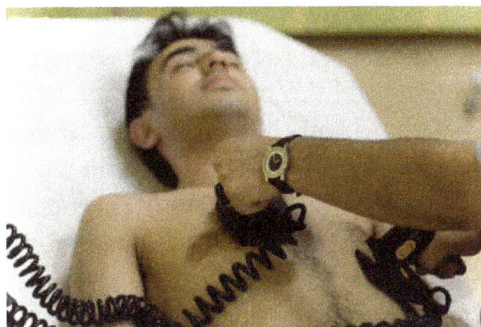
Defibrillator

Sphygmomanometer

The sphygmomanometer is a device used to measure blood pressure (BP), composed of an inflatable cuff to control blood flow and a mercury to measure the pressure. The standard unit of measurement of BP is millimeters of mercury (mmHg) as directly measured with a manual sphygmomanometer. These devices are classified into two types they are Mercury Sphygmomanometers and Aneroid Sphygmomanometers.

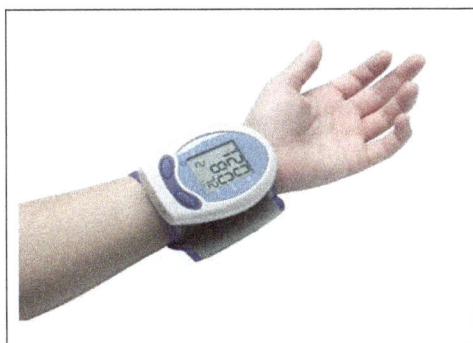
Sphygmomanometer

MRI (Magnetic Resonance Imaging)

The medical resonance imaging technique is used in radiology, to review the natural object of inner elements of the body. They use strong magnetic fields to make pictures of the body. Magnetic resonance imaging includes a big selection of applications in diagnosing and there calculable to be over scanners in use worldwide. Magnetic resonance imaging has an effect on identification and treatment in several specialties, though the effect on improved health outcomes is unsure.

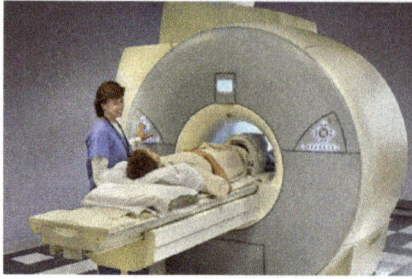

Magnetic resonance imaging

Since magnetic resonance imaging doesn't use any radiation its use is usually recommended in preference to CT once either modality might yield similar information. Image resonance is normally a security technique, however the amount of incidents inflicting patient damage has up. Contraindications to magnetic resonance imaging body most tube-shaped structure implants and cardiac pacemakers, shell gold and foreign bodies within the orbits. The protection of magnetic resonance imaging throughout the primary trimester of physiological condition is unsure; however, it's going to be desirable to different choices. The sustained increase in demand for magnetic resonance imaging among the attention trade has died to considerations concerning price effectiveness and other medical specialty.

Stethoscope

The stethoscope is an audio medical device for listening to the internal sounds of the human body or an animal. It is frequently used to hear the sounds of the lung and heart. It is also used to listen to intestines and blood flow in veins and arteries. It is used In combination with a sphygmomanometer for measurements of blood pressure. Stethoscopes can also be used to check scientific vacuum chambers for leaks, and for various other small-scale acoustic monitoring tasks.

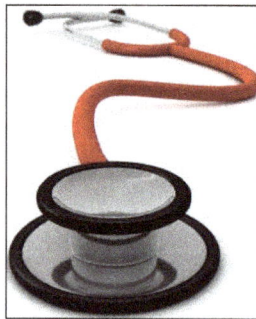

Stethoscope

Therefore, this is all about Medical Electronics Applications in Engineering which includes MRI, Stethoscope, Sphygmomanometer, Defibrillator, IR thermometers, Digital thermometers, Electronic cardiac monitor, Brain wave machine, and Blood glucose monitor.

Automotive Applications

Application of solid-state devices such as diode, silicon-controlled rectifier (SCR), thyristors, gate turn-off thyristors, TRIAC, bipolar junction transistor (BJT), Power MOSFET and so on for control and conversion of electric power is called as power electronics. Application of power electronics in automotive applications plays a major role in controlling automotive electronics. Automotive electronics include modern electric power steering, HEV main inverter, central body control, braking system, seat control, and so on.

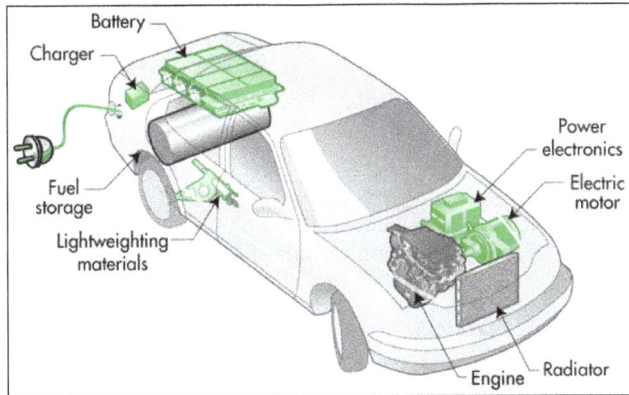

Power electronics in automotive applications

Need of Power Electronics in Automotive Applications

In our day-to-day life, we frequently observe heat radiating from car engine after the car has been driven for a certain distance. This is due to power train system of automotive electronics with an engine or internal combustion or motor as one of the subsystem operating with high temperature exceeding 125 degrees Celsius. Application of power electronics with components such as silicon-based power MOSFETs and IGBTs that are used as power electronic switches in the power train system of automotive electrical and electronic systems for reducing the overall size. And also for managing thermal issues in which a high power of kW range is being used for improving fuel efficiency.

Silicon based dual channel MOSFET

Limitations can be overcome by using a wideband gap semiconductors like silicon carbide with a high-operating temperature that allows placing the circuit near high temperature location. It has two or three times higher thermal conductivity than silicon, which will eliminate need of big copper blocks and water jackets. Silicon carbide has high breakdown voltage and capable of switching at high frequencies with very less power loss which makes the overall size of circuitry very small.

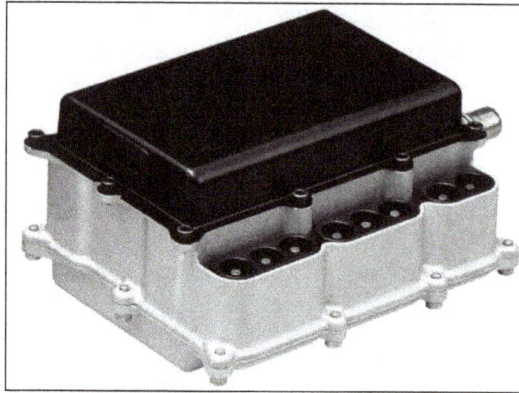

Silicon carbide chip

Application of Power Electronics

Power electronics applications are extended to various fields such as Aerospace, Automotive electrical and electronic systems, commercial, industrial, residential, telecommunication, transportation, utility systems, etc. In case of automotive electronics, the electrically-generated systems are used in automobiles such as road vehicles like telematics, in-car entertainment systems, carputers, and so on. The need to control engines of automobiles originated in automotive electronics for proper controlling and conversion.

Automotive electronics components

Automotive electronics are classified into different types: engine electronics, transmission electronics, chassis electronics, active safety, driver assistance, passenger comfort and entertainment systems. For any power system such as DC/DC or DC/AC or AC/DC, the power electronic components like controllers, gate drivers, converters and so on are required. Generally, based on the vehicle or power supply manufacturer requirements the analog or digital controllers are chosen such that the following parameters including cost, integration, reliability and flexibility are taken into consideration.

Power Electronics Application in Automotive Electronics

Applications of power electronics in automotive electrical and electronic systems includes high voltage systems, automotive power generation, switched mode power supply (SMPS), DC to DC converters, electric drives, traction inverter or DC to AC converter, power electronic component, high temperature requirement, application of SMPS in power train system, and so on. For example, consider a modern car, in which we can find many power electronic components such as ignition switch, control module, vehicle speed sensor, steering sensor and other components, as shown in above figure.

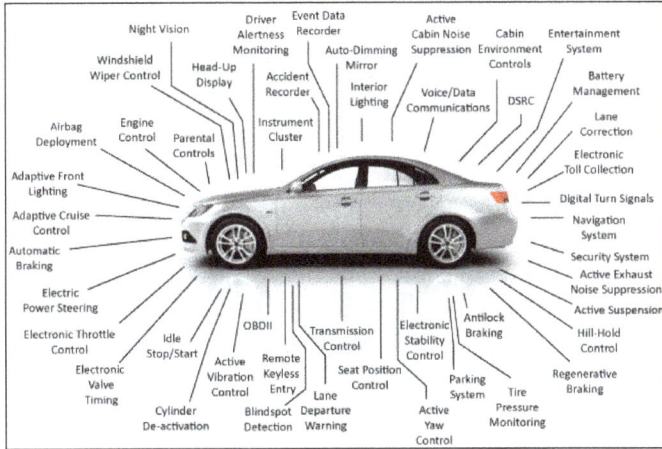

Power electronics application in automotive electronics

Automotive Power Generation

Application of power electronics in the automotive power generation system provides automotive alternators with improved efficiency and high power, along with high temperature withstanding capacity and high-power density with a variety of research in designing of alternator with a switched mode power electronics applications. The frequently used alternator in automotive applications is Lundell or Claw-pole alternator, as it is suitable for the required emerging performance. Field and armature characteristics of this alternator are enhanced by the use of power electronics. These alternators are used in automobiles for supplying power to the batteries and electrical system while the engine is running. Automotive alternators require a power electronic voltage regulator for producing a constant voltage at the battery terminals by modulating small field current.

Cut view of lundell alternator

Switched Mode Power Supply (SMPS)

SMPS concept is based on the power electronics devices such as semiconductor devices that operates in an on state that has zero voltage and an off state that has zero current during this state theoretically with 100% efficiency. To switch these power semiconductor devices on and off the pulse width modulation (PWM) technique is used. Less bulky and small-sized power electronics based converters are used for high frequency switching as these switches are capable of operating under high switching frequencies.

SMPS

SMPS Applications in the Power Train System

The power train systems of HEVs, electric vehicles and ICE need the following SMPS conditioners such as:

- Regenerative braking (AC/DC)

- On-board charger (AC/DC)

- Dual-battery system (DC/DC)

- Traction motor (DC/AC)

DC to DC Converters

There are different DC to DC converter topologies available which can be used based on the requirements. These topologies are classified as isolated and non-isolated topologies which are adopted in power train systems. The application of power electronics in switching has brought a concept of soft-switching where the switches are subjected to low stress using an LLC or resonant mode. These soft-switching, highly reliable and longlife converters are very useful in the automotive electronics market. There are bidirectional converters such as 400 to 12V for electric vehicles and 48 to 12V for hybrid electric vehicle or internal combustion engine.

DC-DC Converter

Traction Inverter (DC/AC)

Electrical motors are machines used for converting electrical energy into mechanical energy and primarily DC motors are used for this purpose, but due to the unreliability of DC motors, AC motors are used because of their efficiency. Application of power electronics in building controllers for AC motors has tremendous progress from the past two decades. Thus, for AC motors to supply power, power stored in batteries of the automotive electrical and electronic systems of electric vehicles or hybrid electric vehicles or ICE require the application of power electronics such as DC to AC converters or electrical inverters.

SPI Inverter

On-board Charger (AC/DC)

Vehicles with automotive electronics consist of batteries that need to be charged; for this charging purpose, the supply AC power has to be converted into DC. We know that, the power can be stored in batteries only in the form of DC. This conversion of AC to DC can be done by the application of power electronics converters called as rectifiers.

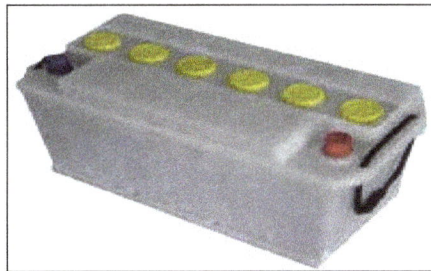

Automotive Batteries

The application of power electronics is increasing with the advancing technologies in automotive electrical and electronics systems for improving the overall system efficiency with high operating temperature, increasing flexibility, reliability and to reduce the overall size of the circuitry.

Applications in Electrical Engineering

Power Electronics is a way of converting electrical energy from one form to another, the output by the conversion is a better, more efficient, error free, clean, compact, simple and convenient to use.

The study of power electronics and electric motor drives involves processing of electric power for a variety of uses. And provide a wide range of application within the range from a few watts to megawatts. For the example we can take various computer disk drives, drills with variable speed, lamp dimmers, electric vehicles, industrial machine tools, and electrical drives.

The area of power electronics is so vast that every day average 12 billion kilowatts per hour of power which is more than 80 % of the power generated is converted or processed or recycled by some of power electronic devices. As the power conversion efficiency is low the losses of energy during the power conversion are so high. We can understand this by a small example of estimated power consumption in a desktop sold in one year, which is equal to the 17 power plants of 500 MW. So it is very necessary to improve the efficiency of the power conversion systems.

By an estimated calculation, it has been found that with the use of efficient and economically effective power electronic technology, the world can save 35 % reduction in energy consumption.

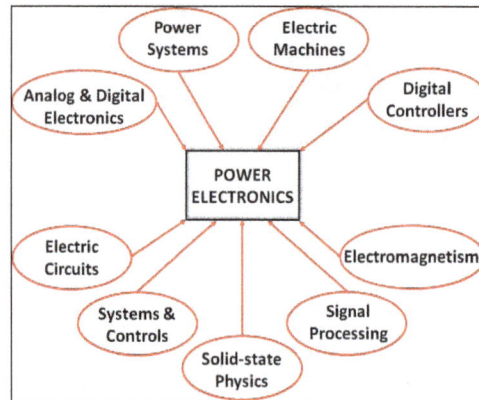

Following are various applications with a wide power range from a few tens of watts to several hundreds of megawatts. As power semiconductor devices improve in performance, efficiency and if the cost will be reduced, more systems will undoubtedly use power electronics.

1. Residential

- Refrigeration and freezers,

- Space heating,

- Air conditioning,

- Cooking,

- Lighting,

- Electronics (PCs, other entertainment equipment.

2. Commercial

- Heating, ventilating, and air conditioning,

- Central refrigeration,

- Lighting,

- Computers and office equipment,

- Uninterruptible power supplies (UPSs),

- Elevators,

- Industrial,

- Pumps,

- Compressors,

- Blowers and fans,

- Machine tools (robots),

- Arc furnaces and induction furnaces,

- Lighting,

- Industrial lasers,

- Induction heating,

- Welding.

3. Transportation

- Traction control of electric vehicles,

- Battery chargers for electric vehicles,

- Electric locomotives,

- Street cars, trolley buses,

- Subways,

- Automotive electronics, including engine controls,

- Utility systems,

- HVDC,

- SVC,

- Supplemental energy sources (wind, photo voltaic), fuel cells,

- Energy storage systems,

- Induced draft fans and boiler feed water pumps,

- Aerospace,

- Space shuttle power supply systems,
- Satellite power systems,
- Aircraft power systems,
- Telecommunications,
- Battery chargers,
- Power supplies (DC and UPS).

It is literally impossible to list all the application of power electronics today; it has covered almost all the areas where electrical energy is being used. This has become a trend now and it is increasing, especially with the present scenario of new devices and integrated design of power semiconductor devices and controllers. The ease of manufacturing has also helped these devices to be available so now a day these devices exist in a vast range of ratings and gradually have appeared in high voltage and extra high voltage systems also. In the end it can be summaries in the words that, the day is not far when all of the electrical energy in the world will pass through power electronic systems.

Permissions

Index

www.ingramcontent.com/pod-product-compliance
Lightning Source LLC
Chambersburg PA
CBHW061257190326
41458CB00011B/3692